COMPLETE TABLE SAW BOOK

BOOK

Revised Edition

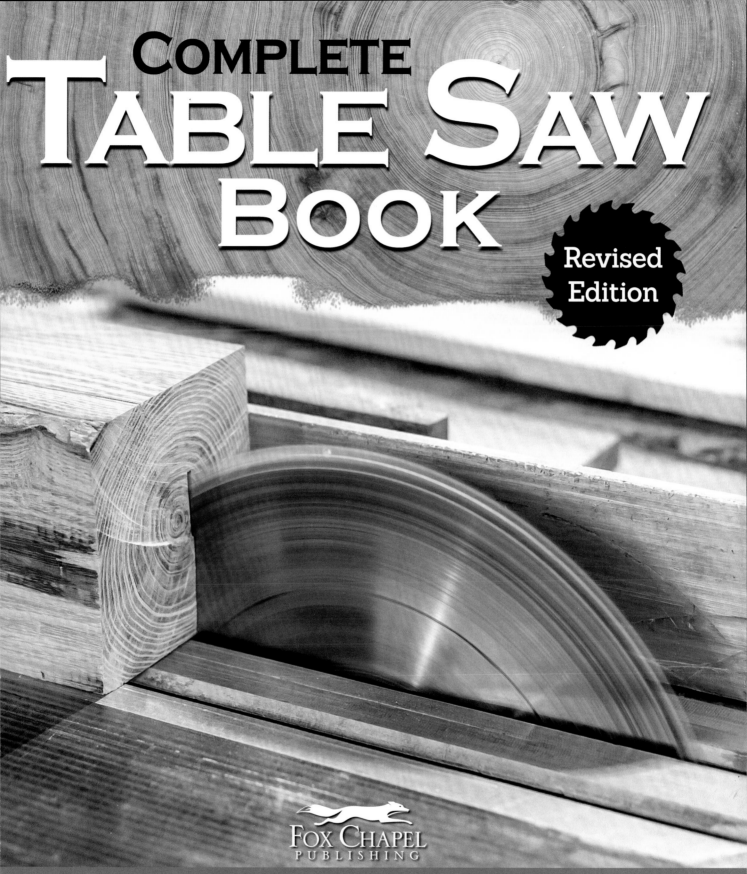

COMPLETE
TABLE SAW
BOOK

Revised Edition

Step-by-Step Illustrated Guide to Essential
Table Saw Skills, Techniques, Tools, and Tips

FOX CHAPEL
PUBLISHING

Chris Marshall

© 2003, 2021 by North American Affinity Clubs

Published by Fox Chapel Publishing Company, Inc., 903 Square Street, Mount Joy, PA 17552.

Fox Chapel Publishing Team:
Editor: Katie Ocasio
Copy Editor: Colleen Dorsey
Designer: Chris Morrison
Indexer: Jay Kreider

Original Edition Team:
Creative Director: Tom Carpenter
Author and Editorial Coordinator: Chris Marshall
Book Products Development Manager: Mark Johanson
Photo Production Coordinator: Dan Cary
Series Design, Art Direction, and Production: Bill Nelson
Lead Photographer: Mark Macemon
Photography: Ralph Karlen
Illustrator: Craig Claeys
Technical Illustrator: John Drigot
Production Assistance: Eric Melzer and John Nadeau
Historical Table Saw Consultation and Imagery: Renier Antiques
Senior Book Development Coordinator: Jennifer Guinea
Contributing Manufacturers: Craftsman Tools, Delta International Machinery Corp., and HTC Products, Inc.

Special thanks to: Renata Mastrofrancesco, Delta International Machinery Corp.; Kelly Mehler, Technical Consultant; Paul Williams, Tried & True Tools, Fridley, MN

ISBN 978-1-4971-0117-3 (paperback)
ISBN 978-1-4971-0202-6 (hardcover)

Library of Congress Cataloging-in-Publication Data

Names: Marshall, Chris, 1967- author.

Title: Complete table saw book / Chris Marshall.

Description: Revised edition. | Mount Joy : Fox Chapel Publishing, [2021] |
    Includes index. | Summary: "Informs readers about table saws, including
    maintenance, blade types, basic cuts, joinery techniques, and
    accessories. Also includes several projects for the workshop and home"--
    Provided by publisher.

Identifiers: LCCN 2020049519 (print) | LCCN 2020049520 (ebook) | ISBN
    9781497101173 (paperback) | ISBN 9781497102026 (hardback) | ISBN
    9781607658368 (ebook)

Subjects: LCSH: Circular saws. | Woodwork--Patterns.

Classification: LCC TT186 .M35 2021 (print) | LCC TT186 (ebook) | DDC
    684/.08334--dc23

LC record available at https://lccn.loc.gov/2020049519

LC ebook record available at https://lccn.loc.gov/2020049520

To learn more about the other great books from Fox Chapel Publishing, or to find a retailer near you, call toll-free 800-457-9112 or visit us at www.FoxChapelPublishing.com.

We are always looking for talented authors. To submit an idea, please send a brief inquiry to acquisitions@foxchapelpublishing.com.

Printed in China
First printing

TABLE OF CONTENTS

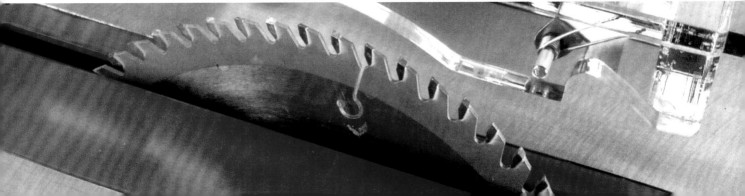

Part 2: Woodworking with Sheet Goods

Part 3: Projects

WELCOME TO COMPLETE TABLE SAW BOOK

The table saw is the workhorse of the woodshop. Cabinetmakers, fine woodworkers, and even carpenters rely on it to perform the cutting operations that are most essential to their jobs. In each of these cases, other tools become important for the finished execution of the project. But without the table saw to rip stock down to size, projects would bog down before they ever left the gate.

Cutting lumber and sheet goods to usable and finished sizes is a task at which the table saw excels, but this versatile stationary tool has much more to offer. Cross-cutting, bevel-cutting, and miter-cutting can be done with accuracy and ease on a table saw. You can also use it to cut a host of wood joints, from basic butt joints to blind dadoes to finger joints. You can even equip a table saw with specialty cutters and shapers for making your own custom moldings.

With all these abilities at your disposal, the universe of projects you can accomplish will expand immeasurably once you put a table saw at the hub of your shop. But a saw is still only a tool. To achieve pleasing results with accuracy and safety requires a lot from the saw's operator, too.

The *Complete Table Saw Book* contains all the information you need to choose and use a table saw effectively. You'll find indispensable tips on basic tune-ups and maintenance; critical information about choosing blades and accessories; step-by-step instructions for making a dozen woodworking joints on a table saw; and straightforward, fully illustrated guidance for making foolproof table saw cuts. We've even thrown in a little table saw history to boot.

Armed with such comprehensive information on the table saw, you'll soon find that you're itching to put your new skills to work. That's why we've included step-by-step plans for nine useful woodworking projects that leverage the table saw's extensive capabilities. Use these detailed plans to build basic cabinets, a desk and console, workshop projects,

and a couple of functional shelving units for your home. For this book, we deliberately chose to show projects that are built entirely (or almost entirely) with sheet goods. Plywood, particleboard, and medium-density fiberboard (MDF) are ideal materials to be cut and shaped on a table saw. No other tool is as adept at taming these heavy sheets of stock. You may not automatically think of sheet goods as being suited for fine woodworking, but once you look at the projects shown here, you'll change your mind.

A table saw is a major investment that consumes valuable floor space in any shop. By taking the time to learn all of the ins and outs of this hardworking tool, you can cause your investment in money and space to pay high dividends. When you use your table saw to build beautiful projects for your home, the dividends will increase even more. The *Complete Table Saw Book* is the only advisor you need to reach this potential. We think it will bring you many happy returns.

Important Notice

For your safety, caution and good judgment should be used when following instructions described in this book. Take into consideration your level of skill and the safety precautions related to the tools and materials shown. The publisher cannot assume responsibility for any damage to property or persons as a result of the misuse of the information provided. Consult your local building department for information on permits, codes, regulations, and laws that may apply to your project.

PART 1
Table Saws
INTRODUCTION

Buying your first table saw is a rite of passage for most aspiring woodworkers. There comes a point in your woodworking pursuits when you reach the cutting limits of your circular saw or jigsaw and you want more: more versatility, more accuracy, more control, more power. A table saw will satisfy these needs and open new doors of possibility as your woodworking skills grow. You may discover that owning a table saw is just the push you've needed to step up to the challenges of more serious woodworking.

If you're a seasoned woodworker already, your table saw is no doubt a reliable companion in the shop. It is certainly one of a handful of tools that gets put through its paces on a regular basis and makes more sawdust than gathers dust. As you've surely discovered, once you've owned a table saw and put it to good use as a cutting, joint-making, and shaping tool, it's hard to imagine doing without one.

Whether you've just unpacked your first table saw or have been using one for years, this book belongs in your shop. Part 1 was developed by table saw users for table saw users. We hope that you'll keep it next to your saw owner's manual as an essential source of no-nonsense information about using, maintaining, and improving your saw. Our goal in creating this handbook is simple: We want you to maximize your saw's full potential in the safest way possible.

In this section, you'll learn the nuts and bolts of what makes a table saw the efficient cutting machine it is designed to be. We'll help you clarify the kinds of questions you should ask when making a decision about which saw to buy, even if it means purchasing a used saw instead of new. Should you decide to buy a used saw, this section outlines what you need to know to conduct a thorough saw inspection. If the saw passes these tests, you should go home with a sound saw that will give you years of good service.

Maybe your saw isn't cutting as accurately as it once did and is overdue for a thorough tune-up. We've dedicated a full chapter to tuning and maintenance, which walks you step-by-step through those essential tweaks and adjustments you should make from time to time to keep your saw in tip-top shape. If a good tune-up doesn't do the trick, you might be able to replace a part or two on your saw to boost its cutting performance. Read the chapter on sawing accessories to learn more about aftermarket safety and precision cutting devices that may be available for your saw. Customizing the tool you already own could save you hundreds of dollars over investing in a new saw.

But all of this information is secondary to what's most important to any woodworker: cutting wood. Using a table saw implies two requirements. First, saw with a working blade guard and splitter or riving knife whenever the cut will allow it. (Note: On some of the photographs in this book, the blade guard and splitter or riving knife have been removed for photographic purposes only.) Second, learn the fundamentals of safe sawing techniques. This section will teach you what you need to know to saw safely. You'll learn about choosing and preparing lumber for the saw, selecting the proper blades for your cutting tasks, setting up your saw and your shop for safe sawing, and mastering essential table saw cuts, including ripping, crosscutting, mitering, and more. We've included a complete chapter on essential table saw jigs you can build to make your saw even more versatile, as well as an exhaustive section on cutting a host of different joints to suit any project.

Finally, if you are a history buff, you'll appreciate studying a variety of different vintage table saws dotted throughout this book in sidebars on table saw history. The engravings and lithographs come from numerous tool catalogs printed over the last century and reveal a rich evolution of table saw style and improvements that have influenced the form and function of the saws we use today.

With hundreds of easy-to-understand full-color photographs and illustrations, we hope this book will be a resource you'll pick up again and again to answer table saw questions, improve your sawing skills, and help make your saw both accurate and a pleasure to use.

TABLE SAW BASICS

Regardless of size, price, or style, every table saw is really just a flat table through which a blade protrudes to cut lumber. A couple of fences—one running parallel to and another perpendicular to the blade—guide workpieces as you push them into the blade. Over the past century, table saws have taken various shapes and sizes, but they have always shared an efficient and time-tested relationship between table, fences, and blade.

These days, table saws fit into three distinct families that span considerable ranges in price, features, and performance: jobsite saws, contractor's saws, and cabinet saws. This chapter will help you understand those anatomical features shared by all saws, then clarify the differences between each saw family.

If you are preparing to buy your first table saw, use these pages to familiarize yourself with the kinds of questions you should ask before you buy. Then try out a few saws to see which features and level of performance suit your needs. Ask questions of tool dealers. Read saw reviews published in reputable woodworking magazines, especially annual tool buyers' guides where saws are ranked side by side. Attend woodworking shows that come to your town and watch table saw demonstrations. Be curious and diligent, and you'll surely find a saw that fits your woodworking or workshop needs perfectly.

On the other hand, if you are already on your second or third table saw, take a few moments to skim this chapter as a good refresher on the basics. It never hurts to reaffirm what you already know.

Finally, study the safety issues outlined at the end of this chapter to learn more about how you can outfit yourself, and your saw, for safe, pleasurable sawing.

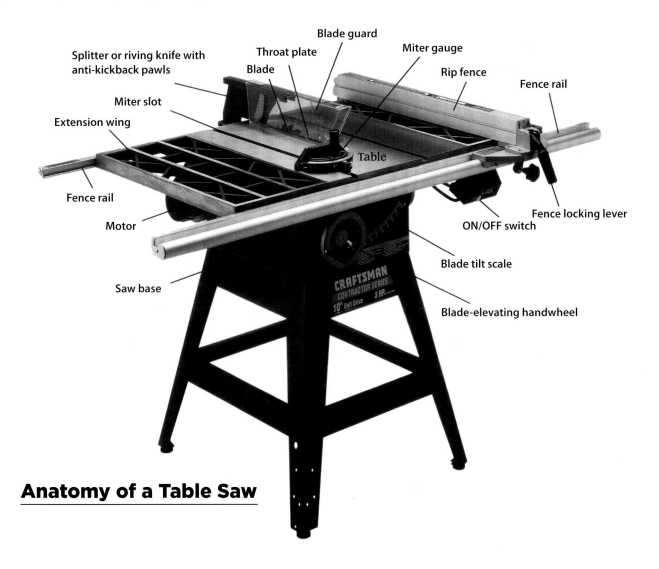

Anatomy of a Table Saw

Saw Table Styles

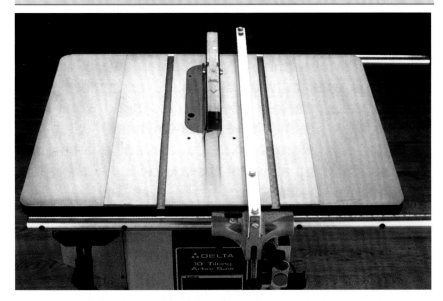

Cabinet saws: Both the center table and extension wings on cabinet saws are made of cast iron, the most durable and flat tabletop material available. All of this iron adds a considerable amount of stability—and weight—to the saw.

Contractor's saws: The center section of the saw table is cast iron, but the extension wings often are made of pressed steel or webbed cast iron (as shown here) instead of solid cast iron. Cast-iron extension wings are preferred over pressed steel.

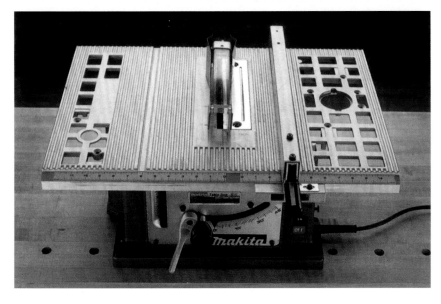

Jobsite saws: The saw table and extension wings on most jobsite saws are made of a single piece of cast aluminum, which is lighter and generally less flat than iron tabletops.

TABLE SAW ANATOMY

Here is a brief overview of the major components found on most table saws.

Saw Table

The center section of most saw tables is made of cast iron to provide a rugged and dead-flat surface. Some saw manufacturers use aluminum alloys for saw tables to cut down on weight, particularly on portable jobsite saws. Saw tables may either be one piece or a combination of a center cast table with two extension wings bolted on each side. Extension wings are made of cast iron or pressed steel, and they enlarge the overall tabletop to support larger workpieces.

Throat Plate

Saw tables are equipped with a removable insert—called a throat plate—that surrounds the saw blade and sits flush with the tabletop in a recessed opening. Essentially, throat plates serve two functions: they support workpieces at the point of contact with the blade, and they provide quick access to the blade arbor for maintenance and blade changes.

Most throat plates do not fasten into the throat plate opening in the table, although some have a metal "nub" that catches underneath the saw table to keep the throat plate from lifting up during a cut. Throat plates typically employ two to four Allen screws so the plate can be adjusted flush with the table surface. Aftermarket throat plates are available for use with dado blades or molding heads. You can also build your own throat plates from hardwood (see page 41).

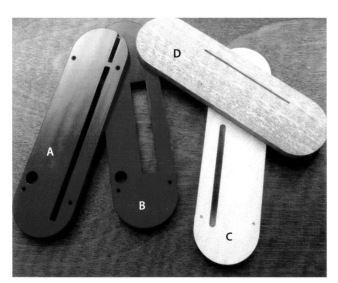

Throat plates that come with most table saws (A) are made of soft cast metal. You can also buy metal dado-blade throat plates (B) and plastic "zero-clearance" throat plate blanks (C). A throat plate does not have to be made of either metal or plastic; hardwood (D) is a suitable alternative and just as safe.

Rip Fence Styles

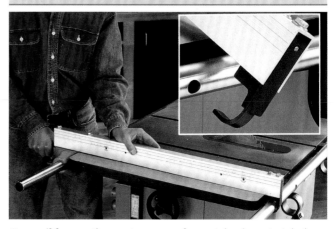

Two-rail fences, the most common fence style, clamp to tubular rails on the front and back of the saw table. A rod extends through the fence body and attaches to a hook-shaped clamp (see inset photo) on the back end of the fence. When you lock down the fence in front, the rod pulls the hook tight in back.

T-square fences, a newer and generally more accurate fence design, clamp to a single hollow rail along the front edge of the saw table. The back of the fence simply rests on a metal bar. Some T-square fences use an adjustable metal button (see inset photo) to glide on the back bar.

Most jobsite saws have light-duty fences that clamp directly to the saw table. A flat plate on the end of the fence (see inset photo) helps to pull the fence tight.

Miter Slots and Rip Fence

You must pass workpieces straight through the blade to cut them accurately. Any deviation from a straight path will lead to sloppy cuts, unsafe sawing conditions, or both. Saw tables are outfitted with an adjustable rip fence and miter gauge to make guiding workpieces easy and safe. Two parallel miter slots machined across the saw table on each side of the blade serve as tracks for the miter gauge.

Rip fences guide boards so they remain parallel to the blade when they are cut lengthwise (called rip-cutting). Depending on the saw, the rip fence body will be made of steel or extruded aluminum and possibly faced with plastic or laminated wood to provide a slippery, smooth bearing surface for guiding workpieces.

Rip fences must be adjustable so workpieces can be rip-cut to different widths. Typically, rip fences slide on one or two fence rails mounted along the front and back edges of the saw table (see right three photos on page 14). Some rails are made of steel or extruded aluminum tubing, while others are simply lengths of "L"-shaped steel bar. Lower-priced saws may even have fence rails that are molded into the saw table (see bottom right photo, page 14).

The front fence rail is equipped with a measurement scale to help index the rip fence a specific distance from the blade. A lever and clamp mounted on the front of the rip fence pulls the fence tight against the saw table and locks it in place.

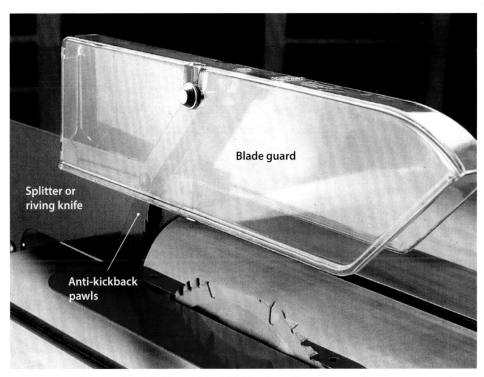

The guard, splitter or riving knife, and anti-kickback pawls protect you from exposure to the blade, as well as workpiece kickback. Standard blade guards attach with a pivoting arm to the splitter or riving knife, so the guard can ride up and over a workpiece, then drop back into place against the saw table.

Guard and Splitter/Riving Knife

Every table saw comes outfitted with a plastic or metal blade guard that shrouds the blade, protecting you from exposure to the blade teeth as well as from sawdust and wood chips that are blown upward by the blade. All table saws have a metal splitter or riving knife that anchors the guard and aligns it with the blade. The primary purpose of a splitter or riving knife is to keep wood from pinching the back (outfeed) side of the blade during rip cuts. Splitters and riving knives are outfitted with a pair of spring-loaded anti-kickback pawls that allow a workpiece to slide in only

> ### Caution
>
> Regardless of how much easier it may be to saw without a blade guard and splitter or riving knife in place, you are completely vulnerable to both cuts and kickback without these safely items. Leave them on your saw for every cut that allows them, or replace them with an aftermarket guard and splitter or riving knife (see pages 140–142).

one direction—from the infeed side to the outfeed side of the blade. In the event that the blade should grab a board and attempt to shoot it back out of the cut (a dangerous condition called kickback), the sharp points on the pawls grab the wood and hold it against the saw table.

Every table saw needs a functional guard and splitter or riving knife to operate safely. However, splitters or riving knives can fall out of alignment and guards can obscure your line of sight when you make cuts, some saw owners remove these safety devices and set them aside permanently.

Standardized Riving Knives Ensure Greater Table Saw Safety

Underwriters Laboratories UL 987 Standard for Stationary and Fixed Power Tools mandates that all table saws designed after 2008 must have a riving knife and guard that rises and falls with the saw blade, instead of a fixed-position splitter and guard. It's a feature that has been standard on table saws in Europe and other countries but had not yet been adopted in North America. With a riving knife in place, the gap between the back of the saw blade and the front, curved edge of the riving knife remains unchanged, regardless of blade height. Formerly, the gap between the blade and splitter would vary—it increased as the blade was lowered and decreased as the blade was raised. Riving knives now help to ensure that a constant and close proximity to the blade will reduce the possibility of workpieces pinching the blade as they exit the cut during rip cut operations and causing a kickback accident.

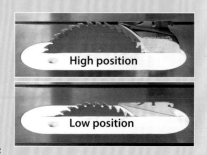

High position

Low position

Miter gauge

A miter gauge is comprised of a swiveling metal head mounted on a flat metal bar that tracks in the miter slots. A miter gauge is used for making crosscuts on either side of the blade. The head functions like a protractor, so it can be set and locked to any angle between 30° and 90°. Most miter gauges are outfitted with adjustable positive stops at 90° and 45°.

When the miter gauge head is set at 90°, the miter slots ensure that a workpiece held against the miter gauge will meet the blade squarely (called *crosscutting*).

Note

Some miter gauges mark 0° to be perpendicular, or square.

Saw Controls

Most table saws are outfitted with two handwheels that are used to adjust the blade. The front handwheel allows you to raise and lower the blade, and the side handwheel controls blade tilt. Worm gears on the ends of the handwheel shafts mesh with teeth on either the front trunnion or the arbor assembly and pivot these parts (see illustration, next page). A few models incorporate both blade tilt and blade height functions into one handwheel. On these saws, the handwheel is switched between tilting and height adjustments by flipping a lever located behind or near the handwheel. In either case, once you've set the blade, you simply tighten a knob in the center hub of each handwheel to lock in the blade settings.

The ON/OFF switch on any saw should be positioned prominently and within easy reach. You should be able to trigger the ON/OFF switch without looking at it. Move the switch box on the saw base if it's in an inconvenient place so you can shut the saw off instinctively and quickly in the event of an emergency.

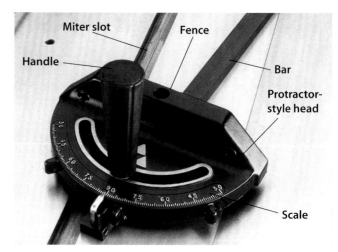

Miter gauges, used for making crosscuts and miter cuts, have a swiveling protractor-style head mounted on a bar that fits into the miter slots in the saw table. A handle screws into the bar and locks the fence.

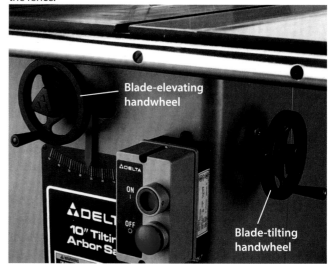

Table saw operating controls should be accessible and sturdy. This cabinet saw features large ON/OFF buttons, an easy-to-read blade tilt scale, and rugged steel handwheels located within easy reach from the front of the saw.

Internal Table Saw Components

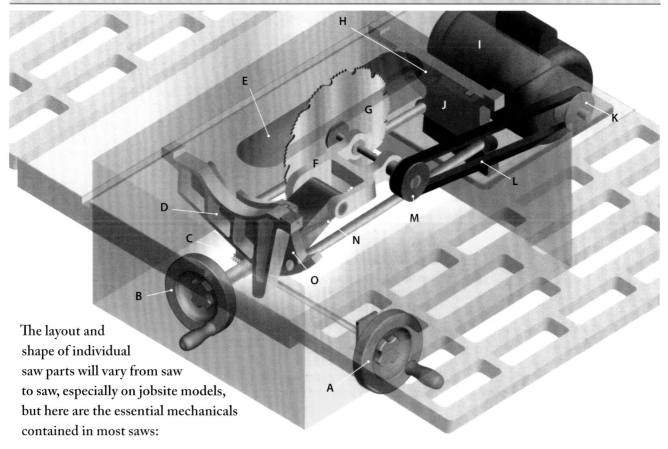

The layout and shape of individual saw parts will vary from saw to saw, especially on jobsite models, but here are the essential mechanicals contained in most saws:

A. Blade-tilting handwheel
B. Blade-elevating handwheel
C. Blade-tilting worm gears
D. Front trunnion
E. Throat plate

F. Arbor assembly
G. Blade arbor
H. Rear trunnion
I. Motor
J. Rear cradle

K. Motor pulley
L. Drive belt
M. Arbor pulley
N. Blade-elevating worm gears
O. Front cradle

Motor and Inner Mechanicals

Table saw "transmissions" consist of an electric motor that sits beneath the saw table or hangs behind it to drive the blade. The blade mounts with a large nut to a shaft called an arbor that spins on two or more sealed bearings set into a cast-iron or steel arbor assembly. (In the case of jobsite table saws, which will be discussed later in this chapter, the blade may bolt directly to the motor shaft, so the motor shaft serves as the arbor.) The motor transfers power to the blade by way of one to three V-shaped flexible drive belts and a couple of pulleys mounted on the motor and arbor shafts.

On cabinet and contractor's saws, the arbor assembly connects to a cast-iron cradle that hangs from semicircular trunnions in front and back. The cradle holds the arbor assembly and motor in line with one another so the drive belts do not have to twist as the arbor tilts. As you turn the blade-tilting handwheel, the cradle meshes with worm gears on the handwheel and slides along arcs formed on the trunnions, which tilts the blade. When you turn the blade-elevating handwheel to raise and lower the blade, worm gears pivot the arbor up and down.

Refer to the illustration above for a general overview of saw mechanicals. The appearance of parts on your saw will vary. To acquaint yourself with the specific mechanicals of your saw, refer to the technical drawings printed in your owner's manual.

Saw Base

All of a table saw's hardware and controls are contained in a sturdy saw base. Saw bases vary in construction, shape, and size, but they provide a stable platform for the saw table, rip fence, and mechanicals. Bases are made of either steel or high-impact plastic. Some bases form floor-standing cabinets (from which cabinet saws get their name), while other bases are shorter in order to sit on a worksurface or are outfitted with legs to make the saw floor-standing.

Jobsite Saws

Jobsite table saws are the smallest table saws you can buy. Most saws in this category have tables that are less than 3' wide and 2' deep, so they can be transported easily. Weighing less than 50 lbs. fully assembled, jobsite saws are easy for most people to carry by themselves.

Given their size and weight, jobsite saws are popular these days with contractors, who can tote them from job site to job site and set them up on the tailgate of a truck or on a couple of sawhorses. A jobsite saw may be a good choice for you if you are an occasional woodworker or weekend remodeler. Those who work in cramped basement shops or who need their full-time garage to double as a workshop will also appreciate the smaller proportions of a jobsite saw. Set it on a shelf or under a workbench, and it's out of the way, without sacrificing any additional floor space.

In addition to being fully portable, consumer-grade jobsite saws are also the least expensive new saws to purchase. Jobsite saws may run on corded 115-volt current or rechargeable lithium-ion batteries. Most jobsite models accept standard 10"-diameter table saw blades, so the small size of the saw doesn't limit your cutting capacity. Jobsite saws can also accept smaller diameter circular saw blades, as can most other table saws with ⅝" arbors.

Along with their advantages, jobsite saws have their limitations. Older jobsite saws may have relatively crude rip fences that are difficult to keep adjusted, steel (rather than cast-iron) components in the undercarriage, and small miter gauges with less precise protractor scales. Saw tables are usually made of aluminum alloys rather than cast iron, so they aren't as rigid or flat as the tables on larger saws. Fortunately, the popularity of these saws is improving the quality of the components that manufacturers use, especially concerning rip fences. Many new jobsite saws are outfitted with rugged rip fences that use rack-and-pinion gears to make them capable of greater precision. But not all jobsite saws are created equal; price is a good indicator of quality.

Jobsite saws are *direct-drive* saws, which means the saw blade mounts directly to the motor shaft. There is no separate arbor shaft. These saws are powered by universal motors, the same motors used in handheld power tools like routers, sanders, circular saws, and corded drills. Universal motors are designed to develop peak power for short periods of time at high revolutions, so they run at relatively high amperage (usually around 15 amps). This means that jobsite saw motors are capable of short bursts of power, which may exceed even 2 h.p., but they can't sustain heavy sawing loads for cutting hardwoods, wet lumber, or long, thick boards.

Larger contractor's and cabinet saws, as you'll see on the following pages, use slower-spinning induction motors that run at lower amperage (230-volt induction motors run at half the amperage of the average jobsite saw). Induction

Jobsite saws are portable, powerful, and inexpensive. Their lightweight construction, however, means they have smaller tabletops, rip fences, and miter gauges and make more noise. They can also produce more vibration than heavier saws. Jobsite saws are best suited for simple woodworking and remodeling tasks.

A jobsite saw without a floor stand or rolling wheelbase can tuck conveniently beneath a workbench or on a shelf, without taking up additional floor space in the shop. Weighing 50 lbs. or less, these saws are easy for most people to lift and carry alone.

motors run cooler and maintain more even horsepower under load. Jobsite saws are powered by universal motors primarily because the motors are much lighter in weight and smaller proportionally than induction motors, so they fit into smaller, more confined saw bases. Universal motors are a perfectly suitable power plant for occasional hard use. However, be aware that these motors are noisier, generate greater vibration, and tend to wear out sooner than induction motors.

Batteries Bring More Versatility to Jobsite Saws

Lithium-ion battery technology has expanded the range of cordless power tool options over the last decade. Advancements in battery capacity and runtime now make it possible to operate some jobsite table saws with 20- to 60-volt cordless tool batteries instead of 115-volt alternating current. A few jobsite saw models can accept either a rechargeable battery or an AC adapter that plugs into household current. While this technology is still limited to portable table saws and not contractor or cabinet saws, the enhancement is particularly helpful to contractors and remodelers who occasionally must operate on jobsites with minimal or no available electrical service.

PHOTO COURTESY OF MILWAUKEE® TOOL

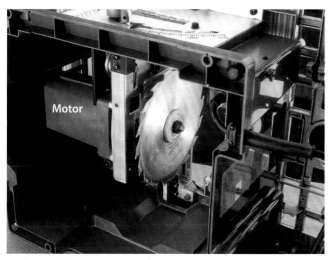

Blades mount directly to the motor shaft on a jobsite saw, just as they do on a portable circular saw. Raising, lowering, and tilting the blade involves moving the motor as well as the blade. There are no drive belts to replace on direct-drive saws.

The layout of blade adjustment controls will vary from saw to saw. This particular jobsite saw uses a knob rather than a handwheel to set blade height. A lever behind the knob controls blade tilt. This saw has scales for both blade height and tilt.

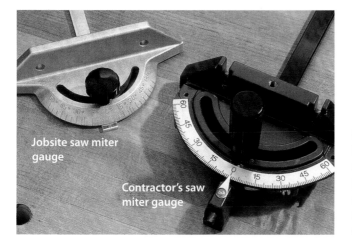

One drawback to jobsite saws is that most are outfitted with lower-quality miter gauges and rip fences that are difficult to set precisely. On the jobsite saw miter gauge shown here, the protractor scale is calibrated in 5° increments, versus single-degree markings on the contractor's saw miter gauge. You'll need single-degree precision from a miter gauge if you plan to do complex woodworking.

Contractor's Saws

If you think about table saws as if they were cars, jobsite saws would be the subcompacts, while contractor's saws would be the mid-priced family sedans. This is because contractor's saws offer the best compromise between cost, quality, and features in a full-sized table saw. Like their jobsite cousins, contractor's saws are designed to be somewhat portable, weighing between 200 lbs. and 300 lbs. But with an overall height of around 34", you wouldn't want to load and unload these saws from a vehicle on a daily basis. Any contractor's saw can be outfitted with a rolling saw base, however (see page 148), so the saw can be rolled wherever it is needed or out of the way for storage.

Contractor's saws are designed for rigorous daily use, whether you are a serious woodworking hobbyist, contractor, or furniture builder. Typical saw table dimensions are around 40" wide and 27" deep, including extension wings, which gives most contractor's saws a ripping capacity (the distance from the blade to the farthest position of the rip fence away from the blade) of nearly 2½'. This table geometry makes contractor's saws capable of ripping full-sized sheets of plywood in half, lengthwise—an unwieldy feat on smaller saws.

One distinctive feature of a contractor's saw is the location of the motor, which hangs outside the saw base in back of the saw. The motor is mounted on a hinged plate, so the weight of the motor actually provides all the tension on the drive belt that spins the blade arbor.

Most contractor's saws are powered by a 1½- to 2-h.p. induction motor, a beefier power plant than a comparable universal motor. Outfitted with a 10"-diameter blade, a contractor's saw can cut through virtually any hardwood or softwood up to 3" thick in one pass. Many saws can be rewired to operate on either 115- or 230-volt current. Converting the motor from one voltage to the other isn't difficult; it's a simple matter of replacing the plug and power cord and reconnecting a few wires (see page 149).

Contractor's saws offer the best compromise between cost, quality, and features in a full-sized table saw. Better-quality contractor's saws have cast-iron arbor and cradle assemblies as well as cast-iron saw tables and extension wings. Precision T-square fences and extension tables now come standard on many models.

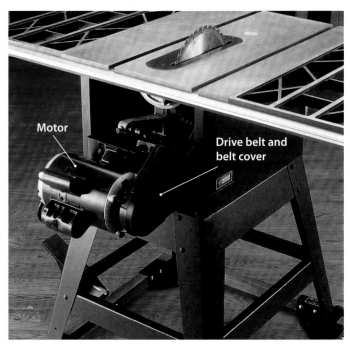

Motor

Drive belt and belt cover

You can immediately identify a contractor's saw if the motor hangs behind the saw base. A drive belt connects a pulley on the motor with a pulley on the blade arbor. Having the motor exposed makes it easier to maintain, but it enlarges the saw's overall footprint.

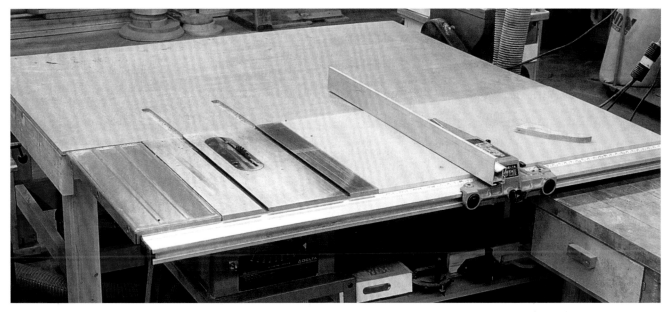

The combination of price and performance makes contractor's saws a lower-cost alternative to cabinet saws for professional woodworkers. This contractor's saw, located in a custom furniture shop, is outfitted with large extension and outfeed tables as well as a precision rip fence. The setup makes it easy for one builder to manage full-sized sheets of plywood.

Contractor's saws are pricier than jobsite saws, but the difference in price is easy to justify when you evaluate performance and features. Undercarriage components like the arbor and cradle assemblies typically are made of cast iron, which means contractor's saws deliver smoother cuts with minimal vibration. Plus, heavier cast-iron undercarriage parts will tend to keep a saw in tune longer than lighter-weight undercarriage parts.

One big advantage to contractor's saws is that they are the target tool for manufacturers of aftermarket add-on equipment. Manufacturers offer a wide assortment of precision rip fence systems, better-quality miter gauges, improved blade guard systems, and accessory sawing attachments like dado blades and molding head cutters. Some contractor's saws come with micro-adjustable rip fences and extension tables as standard items. These accessories can increase the ripping capacity of the saw from 2' to more than 4'.

Aside from a higher price tag than most jobsite saw models, the downsides to owning a contractor's saw are few, but you should consider that contractor's saws unavoidably take up more floor space. Even without long extension tables, a basic contractor's saw will take up about the same amount of floor space as a small riding lawn tractor. Another nuisance involves dust collection. Both the bottom and back of the saw must be covered up to keep sawdust from blowing out of the saw or dropping to the floor. Some saws can be outfitted with a plate that covers the bottom of the saw base and a port for connecting to a shop dust collection system (see page 147).

Table Saw History

Hand-Powering Sawing

Parks Self-Feed Ripsaw (circa 1892)

Prior to the mid 1920s, a saw was either belt-driven from a central power source in the factory or was hand- or foot-powered. Notice the large cast-iron flywheel and handle on the left side of this ripsaw. An operator would spin the flywheel, and the weight of the wheel would keep the saw blade spinning while making a cut.

CABINET SAWS

To continue the car analogy, cabinet saws are the luxury sedans of table saws. They are designed for professional use in commercial cabinet and woodworking shops, but they are also widely available to woodworkers. What sets these saws apart from other table saws in terms of performance and reliability has to do mostly with sheer mass. Cabinet saws feature hefty cast-iron undercarriages and arbor assemblies. Saw tables are made of thicker cast iron, and most cabinet saws also have solid cast-iron extension wings, as opposed to pressed-steel or webbed-iron extension wings found on other saws. To support all of this weight, cabinet saw bases are made of heavier-gauge steel and enclose the motor and arbor assembly on all four sides. Weighing in at more than 400 lbs., cabinet saws aren't portable, though they can be outfitted with rolling bases.

There are two advantages to all of this heavy metal. First, heavier components, particularly the arbor and cradle assemblies and motor mount, dampen vibration. The smoother the saw runs, the less the blade vibrates, which results in smoother cuts. Smoother operation also means that cabinet saws stay in tune for longer periods of time than other saws, especially if they aren't moved around the shop.

The second advantage to heavier components is strength. To meet the demands of commercial woodworking, where they may be cutting thick hardwood or plowing dadoes all day long, most cabinet saws have heavy-duty 3- to 5-h.p. induction motors that run at 230 volts. The typical blade capacity is 10", but saws taking 12"- and 14"-diameter blades are also available for commercial use.

To manage the torque that a larger motor develops, as well as to handle heavy stresses put on the saw blade, cabinet saws often use two or three drive belts to spin the arbor, as opposed to only one drive belt on contractor's saws. Brawny arbor and cradle parts ensure that the motor and belts won't twist the arbor assembly out of alignment. This way, the saw delivers power efficiently from motor to blade without compromising accuracy.

For all this performance and durability, the typical cabinet saw doesn't take up the floor space of a luxury car. Saw table dimensions are about the same as a contractor's saw, not including add-on accessories. Cabinet saws also handle dust collection more efficiently than jobsite or contractor's saws, mainly because the enclosed base captures the dust. New cabinet saws come with a hose port built into the saw base, which can connect directly to a shop dust collection system.

A sizable range of aftermarket accessories are available for cabinet saws, such as precision rip fence and miter gauge systems, improved blade guards and splitters, outfeed support tables, and rolling saw bases. For a sampling of these aftermarket accessories, see pages 136–149.

If all of this performance sounds appealing, brace yourself when you go shopping for a cabinet saw. Heavy-duty construction and professional-quality components don't come cheap. Better-quality cabinet saws start at around two to three times the price of a contractor's saw, and you can spend as much as for a good used car if you buy a professional-grade cabinet saw with all the accessories. Also be aware that you'll need 230-volt service in your shop to power up nearly any cabinet saw, which will probably require hiring an electrician on top of the cost of the saw.

Cabinet saw motors mount to the underside of the saw table inside the cabinet. This helps reduce both noise and dust. Massive cast-iron cradle assemblies provide rock-solid transmission of power between the motor and the blade arbor. The bulk also helps soak up vibration.

Blade arbors on larger cabinet saws are usually made of cast iron, often outfitted with two to four drive pulleys and just as many drive belts. Notice the heft of the cast-iron arbor assembly and trunnions on this saw.

Rugged construction combined with a heavy-duty induction motor make cabinet saws the right choice for rigorous cutting tasks, like working with dense hardwoods or quantities of oversized lumber.

WHICH SAW IS RIGHT FOR YOU?

Despite the fact that there are only three basic saw types to choose from, picking the right saw can still be a tough decision. While the caliber of the saw you use will affect the quality of the cuts you make, spending more money on a professional-grade saw won't necessarily turn you into a better woodworker. Patience, experience, and careful planning always influence how your projects turn out. On the other hand, buying a budget-priced saw that fits your needs today may leave you wanting more from your saw a year or two down the road. So choose a saw wisely, based on your skill level, woodworking expectations, and budget. Here are a few issues to keep in mind when selecting a saw.

Evaluate Your Current Sawing Needs

Consider the kind of woodworking you actually do today to clarify your present sawing needs. Someone who spends an occasional weekend in the shop building small, simple projects is obviously going to have different cutting needs than the person who builds cabinets or custom furniture for a living. If the sum total of your sawing amounts to making butt joints and simple bevel cuts with a circular saw, or if your "woodworking" really boils down to cutting framing lumber for remodeling projects, do you really need a saw with 30" of ripping capacity and a 3-h.p. motor? Probably not. A jobsite saw may be the practical solution for the kind of work you do.

If your woodworking requires a high degree of precision, especially if you frequently cut dadoes or raised panels, buy a saw that offers you smooth, solid performance with an amply powered induction motor of at least 1½ h.p.

Plan Ahead

In addition to evaluating your present needs, be forward-thinking about where your woodworking is headed in the future. If you see your woodworking interests growing, buy a saw that will continue to be a strong performer for years to come. For example, if shop space is at a premium, but you know you'll be cutting dadoes, thick lumber, or larger sheet goods, think twice before you buy a jobsite table saw. Even though the size of the saw might fit perfectly in your present shop space, most jobsite saws are somewhat restrictive in terms of the quality and type of cuts they make. Plus, your accessory options will be limited (see the chart on page 137).

A better solution would be to rearrange your shop (and maybe toss some clutter) to make room for a larger contractor's saw. Think of it this way: When the size of your shop and your skill level grow—and both probably will—the saw will grow with you.

Consider Your Woodworking Budget

Woodworking is no different from any other specialized hobby when it comes to budget-draining expenses. The price range for a table saw is truly wide open. If price and shop floor space were of no consequence, every woodworker would probably own a top-of-the-line cabinet saw. However, most budget-conscious woodworkers would find it difficult to justify paying the huge price of a professional-quality saw. Determine how a saw fits into the rest of the woodworking tool arsenal you still need to buy—routers, planers, jointers, sanders, band saws, power miter saws, drill presses, lathes—the list goes on and on. Then buy as much jobsite, contractor's, or cabinet saw as your budget allows. Don't discount the possibility of buying a used saw, either. For more information on buying a used versus new saw, see pages 25-27.

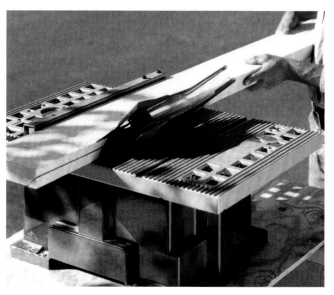

If you spend the majority of your shop time remodeling or building simple woodworking projects, a jobsite saw can provide you the versatility you need at an affordable price.

Table Saw History
Table Saw and Horizontal Borer, All in One

The Egan Co. No. 3 Improved Variety Saw

Around 1902, this belt-driven cast-iron saw served double duty as a horizontal boring machine. As with other saws of its time, the saw table tilted, raised, and lowered, while the saw arbor and blade remained fixed in place. A bit chuck, mounted to the opposite end of the arbor from the blade, was used for boring tasks. A separate drilling table on the right side of the saw could be raised or lowered in relation to the drill bit, then plunged forward into the bit to drill a hole.

The Bottom Line

For most woodworkers, a 10" contractor's saw with a 2-h.p. motor and precision rip fence is a sensible saw to buy. The table is large enough to handle workpieces of all sizes, and the undercarriage and arbor will handle the rigors of cutting with dado blades and molding heads. Since it's a middle-of-the-road saw, you'll be able to customize a contractor's saw as you like, with an assortment of aftermarket accessories.

You'll need a full-sized contractor's or cabinet saw for cutting sheet goods and other long stock safely. Either saw will give you the option of adding other helpful accessories later, like extension tables and precision rip fences.

Multipurpose Machines

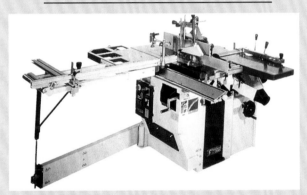

Robland LX Combination Machine

A number of manufacturers offer woodworking machinery that combines several stationary tools in one unit. The table saw is the heart of these machines, configured with other tools like jointers, horizontal borers, planers, shapers, and sanders. Some models have a single heavy-duty motor that is slid into different positions on the machine to power one tool at a time. Others incorporate several motors that work independently to run separate tools. The advantage to owning one of these multipurpose machines is that you can combine a shop's worth of stationary tools in one place. However, multipurpose machines are expensive, and the size of the table and fences for each tool might be limited. Spend some time using one of these machines, then try their individual stationary counterparts to see which approach will fit your woodworking needs best.

BUYING A USED SAW

If purchasing a saw "fresh out of the box" isn't as important for you as getting the most for your money, consider purchasing a used table saw. You'll find used saws advertised in the classified ads of newspapers and woodworking magazines from time to time, as well as at auctions and garage or estate sales. Typically, the market for used table saws is quite good, so be aware that they may be difficult to find and may sell quickly. However, be persistent; aside from the amount of money you could save buying secondhand, there are a couple more compelling reasons to consider a used saw. First, buying used versus new may allow you to get a better-quality saw for the same money. Second, well-cared-for table saws made by reputable manufacturers, especially saws with induction motors and cast-iron tables, arbors, and cradle assemblies, are truly rugged tools with long lifespans. "Used" in no way implies "used up." You may need to replace a few belts or bearings on a 30-year-old saw, but you'll probably still come out ahead price-wise compared to buying a brand-new saw.

A host of hobbyist professional, older, and newer saws are available secondhand, as this used tool store proves. With some careful shopping and a bit of refurbishing, there's no reason a used saw can't perform as reliably and accurately as a new one—and possibly for a fraction of the cost.

Evaluating a Used Saw

Here are some components you'll want to inspect closely when diagnosing the condition of a used saw.

Saw Table

Examine the saw table closely for hairline cracks. They may be difficult to see, and a magnifying glass can help. Cracks are an obvious indicator that the table has been subjected to some sort of trauma. It may have been removed from the saw and accidentally dropped. The saw may have tipped over during transport or could have been used as a makeshift workbench and damaged by a hammer blow. In any case, unless you know a qualified machine shop that can repair the crack, you're better off finding another saw. New saw tables are expensive and can be very difficult to locate for older models that are no longer made.

Carefully check the table for unevenness across the surface using a steel straightedge. It's impossible to do accurate work on a table saw that has an uneven tabletop. Check the saw table for flatness in all directions—lengthwise, widthwise, and diagonally. Note: Inspect how the extension wings are attached to the saw table when you're checking with a straightedge. It's common for the extension wings to be bolted slightly above or below the level of the center table, which will throw off your inspection. You can easily level the extension wings by repositioning or shimming them (see pages 39–40).

Minor warping (1/16" or less across the saw table) can be corrected by having the table ground flat at a machine shop, although this might be expensive and can weaken the table. You can clean up other slight imperfections in the tabletop, like pitting from minor corrosion or scoring from fence and miter gauge wear and tear, by rubbing down the tabletop carefully with rubbing compound or naval jelly and a scrub pad (see page 54). Minor pitting and scoring is nothing to worry about and won't affect the saw's accuracy.

Arbor and Bearings

Another area of concern when considering a used saw is the condition of the arbor and bearings. On good-quality saws with cast-iron and heavy-gauge-steel mechanicals, it takes many years of hard use to generate significant wear, unless the saw was misused or subjected to moisture and allowed to rust.

To test the arbor, remove the throat plate and drive belts that connect the motor to the arbor so the arbor turns freely. Spin the blade by hand and listen (see bottom photo, page 27). If you hear any clicking or grinding sounds, the arbor bearings are worn. Rotate the arbor with your fingers and you'll feel the grittiness of pitted, grinding bearings. If the bearings are worn, they can be replaced or remanufactured by

Use a straightedge to examine the saw table for warping as well as high and low areas. A dead-flat tabletop is critical to saw accuracy. You can correct minor warping (less than 1/16" across the table), but avoid saws with more severe warping.

a good machine shop (they'll be tough to remove and install yourself without specialized bearing pullers and presses).

While you're at it, check the condition of the arbor shaft. It should be smooth and spin without noticeable wobble. Inspect the arbor shaft threads that hold the flange, blade, washer, and nut in place. The threads should be clean and smooth.

> ## Note
>
> In the case of direct-drive saws, the motor armature shaft is machined with threads on the end to serve as the saw's arbor. Use the same arbor tests to determine if the armature bearings are worn. Most high-quality electric motors can be serviced at reasonable prices. The presence of worn armature bearings isn't necessarily a reason to pass the saw up.

Inspect the inner V-faces of the pulleys that connect the drive belt or belts to the arbor and motor. Pulley faces should be clean, flat, and smooth in the area where the belt rides. If the faces bow outward, the pulley or pulleys are worn and the belt will likely slip or burn when the saw is cutting under load. Don't worry if the pulleys are in good shape but the drive belts are cracked or worn; belts are inexpensive to replace. Replace belts with those recommended by the saw manufacturer.

Tip

If the saw has multiple drive belts, be sure to ask for "matched sets" when you order the belts. Matched sets will ensure that the circumferences match exactly so the belts will wear evenly.

Cradle, Trunnions, and Worm Gears

The cradle, trunnions, and worm gears are responsible for supporting and moving the blade arbor. Since these parts move, eventually they will begin to wear. Here's how to evaluate their condition. Lay the saw on its side (or peer down through the throat plate opening). Inspect the condition of the handwheel worm gears and the gear teeth on the trunnions and arbor assembly. Look for uneven wear from tooth to tooth or metal dust around these parts.

Check the arbor bearings for wear by removing the drive belt(s) (on contractor's and cabinet saws) and spinning the blade by hand. Listen for clicking and grinding, which are sure signs of bearing wear. Arbor bearings are fairly inexpensive to replace.

Check the condition of the worm gears and trunnions by cranking both handwheels over the full range of blade movement. The handwheels should turn smoothly, without binding.

Stresses on these meshing parts, coupled with debris and grit that get thrown off the saw blade, can cause gears to wear or chip, especially if they aren't lubricated. (For more on periodic lubrication, see page 54.) Test the action of the gears by turning the handwheels. The gears should mesh evenly, quietly, and smoothly.

Other General Considerations

Take a step back and look at the overall condition of a used saw. Inspect for signs of good saw maintenance and proper usage, including:

- minimal corrosion on metal parts
- the throat plate does not show signs of saw blade damage, an indicator that the saw has been severely out of tune or suffered kickback or some other severe blade trauma
- power cord and plug are free of cracks or frays
- rip fence slides smoothly and completely along the fence rails
- the saw isn't missing bolts, screws, and other fasteners and these parts are tight

Ask the owner for the owner's manual that came with the saw, particularly if the saw is old. You might need it for ordering replacement parts later on.

Test Drive before You Buy

Finally, take the saw for a "test drive." The motor should start without hesitation, bring the blade up to speed quickly, and have an even-sounding pitch when it reaches full power. If you can, make a few test cuts. Does the motor labor excessively under loads? Does the saw vibrate noticeably when it's running or cutting? Does the saw cut cleanly and evenly or does it leave ragged, uneven cuts and burn marks?

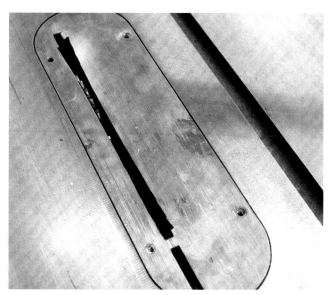

Damaged throat plates are a telltale sign that the saw was abused or severely out of tune at some point. Scrutinize a saw with throat plate damage for other signs of neglect.

Many aspects of saw performance can be improved by simply tuning up the tool or installing a sharp blade, but consider that poor saw performance may indicate a lack of good general care the saw has been given by its owner over time.

Table Saw History

Guard and Splitter Assembly

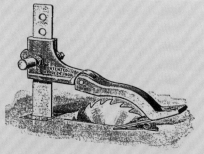

J. A. Fay & Co. guard and splitter assembly (circa 1900)

This guard and splitter assembly, patented by J. A. Fay & Co. in 1900, represents one of the first examples of table saw safety equipment available commercially. Notice that the guard covers the blade from above but not from the side, and it pivots up and down to ride on top of workpieces as they are passed through the blade. The guard, which undoubtedly was made of cast iron and steel, fastened to the splitter with a removable pin. Several holes on the splitter allowed the guard to be raised to accommodate thicker stock.

SAFETY

Table saws are wonderfully efficient and versatile tools, and they're capable of making nearly every straight cut you'll ever need for woodworking, but they have an inherent dark side. Regardless of how carefully you follow safety precautions, you can't change a table saw's basic characteristics. A standard table saw spins a 10"-diameter saw blade at around 4000 rpm, or about 120 mph. Every time you make a cut, the blade puts your hands and forearms at risk of injury, even when all of the saw's safety equipment is installed. This kind of blade velocity also generates a considerable amount of dust and noise. Sawdust is irritating, especially if you have allergies, and the blade noise can be permanently debilitating over time.

The good news is you can easily take measures to protect yourself. The following section covers personal safety items you should have on hand, proper workshop attire for sawing, and other information related to table saw safety. Refer to the next chapter on "Making Basic Cuts," pages 63–103, for more information about those safety precautions you should take as you prepare to make and then execute specific kinds of cuts.

Use Pushing Tools and Featherboards

As a rule, keep your hands as far from the spinning blade as you can without relinquishing control of the workpiece. Here are several simple devices for cutting safely:

A *push stick* is simply a straight or curved handle, usually made of scrap wood, with a notch on the end for holding workpieces securely against the saw table.

Push pads, commonly used with jointers, are rectangular plastic blocks with a handle on top and a non-skid lining. They spread workpiece control over a broader area when pushing boards facedown over the saw.

Pushing tools are inexpensive, but they are also easy to make. There's no specific formula for designing the best push sticks, just a few general rules: you should be able to grip them comfortably; they should keep your hands far enough above the blade to be safely clear of it; and they should allow you to control the workpiece without feeling awkward.

Featherboards hold a workpiece snugly against the saw table and rip fence. They typically are made from a length of hardwood stock with a series of parallel 3"- to 5"-long saw kerfs cut into one edge, forming springy "feathers" in the wood. You can make them easily in the shop (see "Making Featherboards," next page), or you can buy inexpensive wood, plastic, or magnetic-soled models.

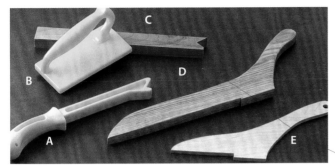

Pushing tools for table saws. Plastic push sticks (A) and push pads (B) are inexpensive to buy, but simple wooden push sticks (C) are easy to make. Keep a long-soled pusher (D) handy for providing broad support as well as a thin pusher (E) for squeezing into narrow spaces between the blade and rip fence.

Making Featherboards

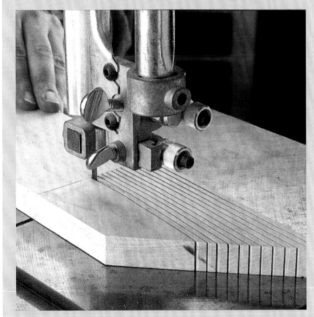

Select a length of hardwood scrap about 3 to 4" wide and 14" to 18" long. The length of the featherboard will depend on how far the featherboard must reach to the edge of the saw table for clamping purposes. Draw a line widthwise across the face of the board, about 5" from one end. Miter-cut the end of the board closest to this layout line at about a 40° angle (the angle can vary anywhere from 30° to 45°). Then, using a band saw or jigsaw, make a series of parallel cuts from the angled end of the board to your layout line to form the flexible "feathers." Space the cuts evenly about ⅛" to ¼" apart.

Featherboards keep workpieces moving in the only safe direction: from front to back through the blade. Some featherboard designs incorporate a powerful magnet (A) to hold them tightly to the table. Other styles clamp to the rip fence (B), to the saw table (C), or pressure-fit by way of a short bar fitted into the miter slots (D).

Keep Your Senses Sharp

Never operate a table saw while under the influence of alcohol or other drugs that make you drowsy, dull your senses, or impair clear thought. It takes only a fraction of a second for sawing accidents to happen, so you have to focus your attention on what you are doing every time you make a cut. Think through the sequence of steps that will be required to perform a cutting operation before you begin, and keep in mind what your safe response should be when you encounter a problem (see the chart on page 74). Stop your work at the table saw if you find yourself starting to lose track of what you are doing or begin to feel frustrated or fatigued. Be particularly careful when you are making repetitive cuts; monotony leads to distraction, which invites accidents.

Dress Appropriately

Wear comfortable, rubber-soled shoes that give you good traction on your shop floor. Roll up long sleeves if they fit loosely, to keep them from getting caught on a workpiece, or worse, the saw blade. Keep pencils and other marking tools out of shirt pockets when sawing so they don't fall out and roll into the blade as you lean over the saw. Do not wear gloves while operating a table saw. You'll get a better grip on a workpiece if you manipulate it bare-handed, and your fingertips will warn you instantly if your feed rate changes or the board begins to behave erratically. Working without gloves means you may have to heat your shop in the wintertime to warm your hands, but extra warmth will keep the rest of your body's reflexes sharper and make your woodworking more comfortable anyway.

Unplug the Saw Whenever It Is Not Being Used

Simply switching the machine off isn't an adequate safety measure, especially when you are working on or near the blade and drive belt(s). Always unplug the saw for added insurance. It takes only a second to do, and it ensures that the saw cannot start again until you want it to.

Some saws are outfitted with "remove-to-lock" keys that fit over a toggle-style ON/OFF switch. When the switch is in the off position, you can remove the key, making it impossible to flip the toggle switch back on. Remove and store this key in a safe place when you are not working in the shop to keep others—especially children—from starting the saw.

Use a Heavy-Duty Extension Cord Plugged into a Grounded Outlet

Since table saws are mostly metal, be sure you choose a properly grounded three-prong outlet for powering the saw, and check the amperage of the circuit that supplies power to the outlet you'll use. Ideally, workshops should be wired with dedicated 20-amp circuits for 115-volt outlets and 40-amp circuits for 230-volt receptacles. Otherwise, you'll likely trip the circuit when using the saw under full load.

Using an extension cord to power your saw can starve the motor of the amperage it requires to run properly, particularly if the extension cord is long or of too small a wire gauge. This condition, called *voltage drop*, results in the motor running slower and hotter than it should, which can lead to premature wear.

When you must use an extension cord, guard against voltage drop by selecting the shortest length of cord that will reach in a gauge heavy enough to provide at least the amperage your saw motor requires. Heavier-than-necessary gauges of extension cord are even better. Use the extension cord chart on this page as a guide to choosing the proper cord, and refer to your owner's manual to determine the motor amperage of your saw.

Pull out the "remove-to-lock" key every time you are finished using your saw. Without the key installed, the ON/OFF switch is locked, and you cannot turn on the machine. Keep the key in a safe place out of the reach of children.

Extension Cord Ratings

To make certain that your table saw runs safely and at peak performance, check the amperage rating for your saw in your owner's manual. Use this chart as a guide for selecting the proper extension cord.

Cord Length	Gauge	Maximum Amps
25'	18	10
25'	16	13
25'	14	15
50'	16	10
50'	14	15

Table Saw History

Gasoline-Powered Table Saw

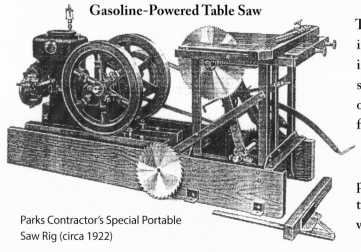

Parks Contractor's Special Portable Saw Rig (circa 1922)

The rise of internal combustion engines in the 1920s even influenced the table saw industry. For a time, Parks offered their swing-style ripping and crosscut saw with an optional gasoline engine. No doubt intended for carpenters, this evolution toward self-powered table saws was a significant step away from either hand- or belt-driven power sources. The machine was advertised to be portable, despite its overall weight of a whopping 850 lbs.

Better, Safer Table Saws from SawStop

While proper training and the use of blade guards reduce the risk of table saw mishaps, these safeguards cannot eliminate the possibility of injuries stemming from blade contact.

In an effort to bring a greater measure of safety to an inherently dangerous machine, SawStop Corporation engineered a proprietary blade brake technology into its line of cabinet, contractor, and jobsite table saws. Every product in this line of unique saws comes equipped with an onboard computer system that senses a person's skin if it touches the spinning blade. The sensor watches the minuscule electric current running through the blade, which changes when it comes into contact with the moisture in a person's skin. When that happens, in less than five milliseconds, an aluminum brake beneath the saw table slams into the blade and stops it. In the same instant, the blade is also lowered beneath the table and power to the motor is shut off.

Once the brake has been activated and has embedded itself into the blade, both the brake and blade must be replaced. However, resetting the saw for use again can be done quickly and easily by the saw owner without the need for special tools or skills. To prevent a false brake trigger, the saw allows you to disable the sensor if you're cutting wet wood, but this means you also lose the safety feature. However, you can test wood before cutting to see if the sensor will engage the safety system.

This skin-sensing injury mitigation technology, unique to SawStop table saws, continues to save many woodworkers and other table saw users from devastating hand injuries. This revolutionary technology is a must-have for anyone who plans to invest in a new table saw.

A view of the blade and brake from under the table.

A close-up of the blade and brake.

Additional Safety Features

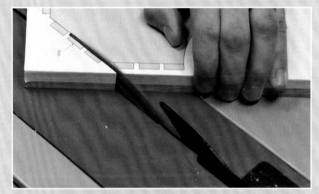

A riving knife helps prevent kickback.

A solid, accurate fence ensures safety and precision.

A push stick is included to keep hands away from the blade.

Jobsite Saw Pro, 1.5 HP
Collapsible rolling cart with small footprint for storage (26-½" W x 29" D x 45" H)

Professional Cabinet Saw, 1.75–3 HP
Cast iron extension wings, dust collection blade guard, blade spacing adjustment gauge

Contractor Saw, 1.75 HP
Steel extension wings, blade spacing adjustment gauge

Industrial Cabinet Saw, 3–7.5 HP
Cast iron extension wings, dust collection blade guard, blade spacing adjustment gauge

Exploring the Jobsite Saw Pro

The jobsite model is SawStop's most budget-friendly option for the average woodworker. Here's a breakdown of its standout features:

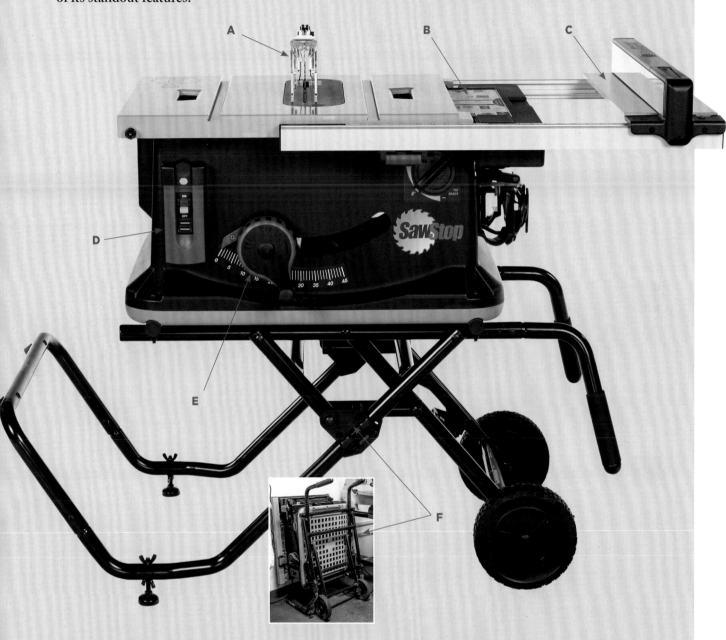

A **SawStop technology instantly drops and brakes the blade if it contacts skin.**

B **Onboard storage for the splitter, saw cover, miter guide, and blade-changing and brake-unit-changing tools.**

C **Extended table with a solid fence makes it easy to rip plywood and other sheet goods.**

D **Oversized power switch doubles as a moisture sensor to prevent the brake from engaging when you cut wet wood.**

E **Quick-adjust blade height and angle guide allow for fast saw adjustments.**

F **Portable saw folds easily but remains solid when unfolded for use.**

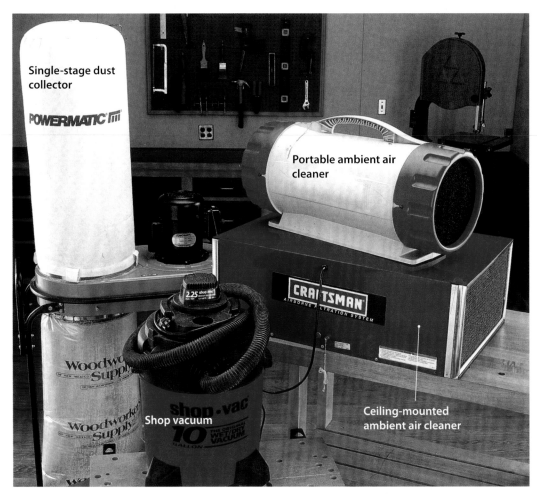

Powder dust collection devices are available for improving air quality in the shop. Single-stage dust collectors attach to stationary power tools through a system of flexible ductwork. A motor-driven impeller draws air through the ductwork and into replaceable collection bags. The bottom bag collects heavier sawdust and chips, while the bag on top filters out finer dust. Ambient air cleaners use synthetic filters to trap fine, airborne dust that would escape a standard dust collector. Both portable and ceiling-mounted styles are available. A shop vacuum is handy for cleaning up excess sawdust in and around the saw.

Protect Your Lungs

Table saws generate large amounts of splinters and chips and clouds of fine sawdust, especially when they aren't connected to a dust collection system or are operated without the blade guard. Studies now indicate that sawdust can be a harmful irritant to mucous membranes in your nose and lungs, especially if you suffer from allergies or other respiratory problems. The dust from some wood species, such as western red cedar, cocobolo, and treated lumber can be particularly irritating or even hazardous to your health in concentrated amounts.

All of this is to say, protect your lungs from sawdust. If you use your saw frequently or in enclosed spaces, it's worth investing in a shop dust collection system. Single- and two-stage dust collectors (see above photo) use a blower and a system of hoses and collection bags to draw dust and debris away from shop tools. Ambient air cleaners are freestanding

or mount to the shop ceiling and draw air through a series of synthetic filters to remove fine, airborne sawdust.

If you choose not to install a dust collection system, wear a dust mask. Many styles of dust masks are available and are rated safe for use in dusty conditions. The least expensive are the disposable paper particle masks. Choose the type with two straps rather than one to get a better seal around your nose and mouth. The more expensive, but better, mask varieties are the respirator-style silicone masks that use replaceable filter cartridges.

If you opt for a dust mask and not a dedicated dust collection system, improve the air quality in the shop by providing for good ventilation. Open doors and windows or saw outside if the weather permits and your saw is portable. Set a fan in a window facing out to help draw dust out. The more fresh air that enters the shop, the better.

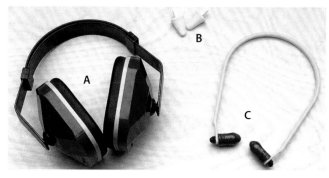

When sawing, wear hearing protection with a noise reduction rating of at least 25 decibels. Choose the most comfortable style for you: earmuffs (A), foam ear inserts (B), or ear bands (C).

Protect Your Eyes and Ears

Flying sawdust and splinters are hazardous to your eyes as well as your lungs. Not only can sawdust scratch your eyes, but it also impairs your vision when you are most vulnerable to bodily injury—in the middle of making a saw cut. Always wear eye protection when operating your table saw.

Prescription glasses are not safety glasses unless they have high-impact polycarbonate lenses. Even then, your eyes aren't fully protected from the top and sides. Choose a pair of safety glasses or goggles with anti-fog, anti-scratch, and anti-static features, and that shield your eyes from all sides. Pick a style that fits your face comfortably, accommodates your prescription glasses if necessary, and hangs conveniently from a safety strap around your neck. The more comfortable and convenient your safety eyewear is, the more likely you'll be to wear it.

Table saws produce about 100 decibels of high-frequency noise, the kind of noise most likely to cause hearing loss. Hearing damage begins when your ears are subjected to noise frequencies higher than 85 decibels for extended periods of time. As with eye protection, do not underestimate the importance of adequate hearing

If you don't install a dedicated dust collection system, wear a dust mask instead, especially if you suffer from respiratory problems. Disposable particle masks (A) provide some protection, but a mask with replaceable filters (B) is even better. A cartridge-style respirator (C) is your best line of defense against dust.

protection in the shop, even if you use your saw only occasionally. Pick a style of hearing protection most comfortable for you. You'll find three essential types: earmuffs, ear inserts, and ear bands. Earmuffs completely cover your ears and wrap around the top of your head. Ear inserts are small foam cylinders, either loose or attached to one another with a length of plastic cord, that you roll between your fingers and insert into your ear canals. The foam expands to block out noise. Ear bands are made of flexible plastic and are outfitted on either end with a plastic cone or foam pad that inserts into or covers your ear canals. The band wraps around the back of your head or hangs under your chin to help hold the earplugs in place. Any of these three hearing protection types you choose will block out the harmful high-frequency noise, but still allow you to hear ordinary conversation. Be sure that the ear protection you wear for sawing has a noise reduction rating (NRR) of at least 25 decibels.

Prepare for Emergencies

Install a first-aid kit in the shop within clear view. Should a table saw accident occur, you'll need a first-aid kit suitable for treating lacerations, amputations, and eye injuries. Learn the proper techniques for stopping bleeding, treating eye injuries, and handling the kind of shock associated with severe injury. If you don't have wall phone or cordless phone in your shop, be sure to keep your cell phone close by in case of an emergency. Let someone know you're working in the shop and have them check up on you from time to time, especially if your shop occupies an outbuilding out of earshot. Finally, equip your shop with a fire extinguisher rated to handle both electrical and combustible fires.

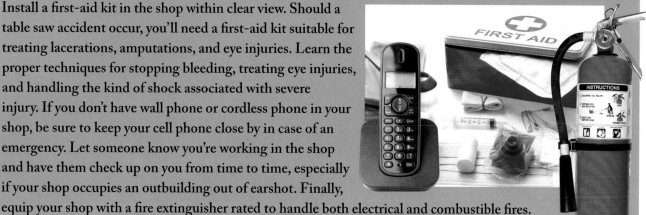

TUNE-UP AND MAINTENANCE

Most woodworkers would rather be cutting boards than taking time out to tune up and maintain their table saws. While it's true that adjusting your rip fence isn't as much fun as ripping a piece of cherry accurately and safely, you really can't have one without the other. That is, don't expect precise and pleasurable cutting from a saw that needs a tune-up.

If you're not the type who enjoys fiddling with machines, the good news is that a number of the tune-up procedures outlined in this chapter only need to be done once when your saw is new and then infrequently after that. The tuning chores you'll need to do more frequently can be taken care of in about an hour.

If you are uncrating and assembling your saw for the first time, don't assume the saw comes tuned up straight from the factory. No saw does. The saw's mechanical components, especially the motor and hand controls, were probably tested before the saw was packaged, but the saw probably wasn't fully assembled, tuned, and then disassembled. The owner's manual that comes with your new saw will provide instructions for putting the saw together correctly, but few manufacturers take you through the next important step of fine-tuning your saw.

After cutting wood for a while, all saws will eventually fall out of adjustment after cutting wood for a while and require a tune-up. It's unavoidable. Tune-ups are required most frequently on lower-quality saws, heavily used saws, and saws that get moved around the shop often. It doesn't take much jostling to throw a splitter or rip fence out of alignment. It's also a good idea to check the tuning on a saw you haven't used for some time or are using as a guest. Treat an unfamiliar saw as though it has never been tuned. You'll safeguard yourself and ensure more accurate cutting right from the start.

Follow the procedures outlined in this chapter to give your saw a thorough going-over before you start that first woodworking project. You'll most certainly get better cutting results.

BEFORE YOU TUNE-UP YOUR TABLE SAW

Learning to recognize when it's time for a tune-up is important in the maintenance of your table saw. It's much easier to do if you learn to recognize the signs and have a plan in place.

Watch for Tune-Up Clues

A saw will exhibit several clues if it's due for a tune-up. Watch for changes on the cut edges of your workpieces. Look for increasing numbers of circular blade swirls or burn marks on the wood. Also, check a workpiece after you rip it to see if the long edges are exactly parallel. At the same time, check to see that the edges are square to the board faces. They should be. If not, make some more test cuts and inspect the results. Consistently sloppy cuts indicate that your saw isn't cutting as it should.

Another possible indicator that your saw needs tuning is if the blade smokes or stalls mid-cut. Smoke indicates that the blade is generating excessive heat and burning the wood. You need to fix this problem to keep from taxing the saw motor, dulling or warping the blade, scorching your workpiece, and as a safety precaution to prevent kickback.

Telltale Signs That It's Time for a Tune-Up

Increasing numbers of burn marks on workpieces is a sign that the blade and rip fence are out of alignment or the blade is dirty or dull. Be aware that some kinds of woods like cherry are prone to burns when you cut them, even if your saw is in tune.

Draw a line along a workpiece and rip-cut the board along your layout line. Does the blade follow your line? If it begins to drift, your blade could be heeling (see pages 44–45) or the fence needs to be adjusted so it's parallel to the blade.

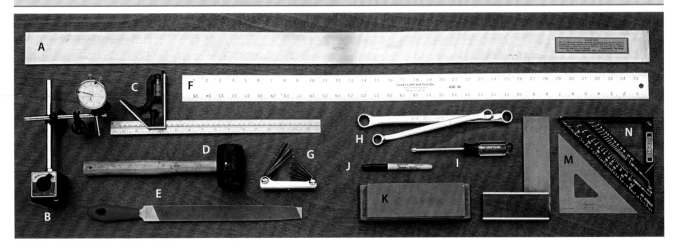

A basic tool kit for tune-ups includes: (A) 4' to 6' steel straightedge; (B) dial indicator with magnetic base; (C) combination square; (D) soft-faced or rubber mallet; (E) flat file; (F) 3' straightedge; (G) Allen wrenches; (H) standard and metric wrenches; (I) screwdrivers; (J) felt-tipped marker; (K) fine-grit sharpening stone; (L) try square; (M) drafting square; (N) speed square.

Once you become comfortable using your saw, you'll acquire a sixth sense for noticing when cutting just isn't right. Trust your instincts. The more wood you cut, the more accustomed you'll become to the particular sound the blade makes as you cut different kinds of wood. You'll also get a "feel" for the right speed to push wood through the blade (called *feed rate*). If the saw motor begins to bog down in a type of wood you cut regularly, a quick tune-up or a fresh, sharp blade likely will solve the problem. A tune-up is also in order if you notice the saw vibrating more than usual or making more noise.

By all means, inspect your saw carefully and tune it up if you experience kickback (see page 77). Kickback indicates one of three possible problems: you are sawing unsafely; your lumber is defective and unsuitable for the table saw; or the blade and rip fence aren't in proper alignment.

Before you cut another board, evaluate what's unusual about what you were doing when the kickback occurred. If you are confident that your cutting techniques are correct, and the troubles you are having aren't due to defective lumber (see page 65), it's time for a saw tune-up.

Tune-Up Goals

Tuning a table saw boils down to correcting four simple geometric relationships between the miter slots, blade, fences, and saw table. On a properly tuned saw, first, the table will be flat across its entire surface, including the extension wings. Second, the blade and rip fence must both be parallel to the miter slots. When this is true, the blade and rip fence will be parallel to one another as well. Third, the blade, rip fence, and miter gauge fence should be square (perpendicular)

to the saw table. A fourth goal is to minimize sources of saw vibration. The usual culprits for vibration are a dirty or dull blade, excessive blade and arbor runout, or misaligned arbor and motor pulleys. If your saw still cuts poorly after satisfying these four goals, you may be using an underpowered saw for the task at hand.

Tune-Up Checklist

How frequently you tune your saw will depend on many variables, including the quality of your saw and how often and rigorously you use it. Refer to the following tune-up chart and timelines as a general guide:

Clean the saw	After each use
Adjust the rip fence to the miter slots	Periodically
Adjust the splitter and anti-kickback pawls	Periodically
Inspect and remedy other sources of vibration	As needed
Set the blade tilt stops	Monthly
Tune the miter gauge	Monthly
Clean saw blades and protect from rust	Monthly
Flatten the tabletop and extension wings	Annually
Align the blade to the miter slots	Annually
Align the motor and arbor pulleys	Annually

FLATTEN THE TABLETOP

Since many of the tune-up procedures explained in this chapter involve the saw table in one way or another, it makes sense to start your general saw tune-up here. Unless the saw table is flat, workpieces won't rest evenly and in turn won't meet the blade squarely when you cut them.

Start your tabletop tune-up by examining the center cast-metal portion of the saw table for warp (also called *wind*). To do this, you'll use a simple set of devices called *winding sticks*, which are nothing more than two flat hardwood sticks about 18" to 20" long, with widths that match. Set the sticks on-edge along opposite edges of the center saw table. Position yourself about 1' from the stick closest to you and adjust your line of sight across the top edge of this stick, looking beyond it to the opposite stick. If the saw table is warped, the top edges of the far stick won't line up evenly with the top edges of the close stick. Check the center table with winding sticks both widthwise and depthwise (see the illustration, below).

A warped cast-iron saw table usually indicates that the metal was not allowed to cure properly when the table was manufactured. If your saw is new and under warranty, return the saw. Otherwise, you may be able to correct minor warp by inserting metal shims at one or more corners where the saw table bolts to the base. Shimming is most effective if the center table is made of a softer metal alloy, such as those found on many jobsite saws. Your other alternative is to have the saw table ground flat at a machine shop—a much pricier fix.

The tabletop tune-up tasks will require using a perfectly flat straightedge about 4' to 6' long. A precision-ground steel straightedge is best, but a good one will cost more than $100. A much cheaper but sufficiently reliable alternative is to make a straightedge instead. Build one from a dimensionally stable material like MDF, solid-core plywood, or seasoned hardwood, and flatten the edges on a jointer.

To check the table, drag your long straightedge on-edge in all directions, checking for dips or rises in the surface. Check carefully; these high and low spots will be slight. You'll be able to see any imperfections easily by shining a light behind the straightedge and looking for slivers of light that appear between the edge of the straightedge and the table surface. Circle high areas on the tabletop with a marker or grease pencil and use a fine-grit sharpening stone or wet-dry sandpaper to flatten them out. The goal here is to just even out high and low areas, so proceed carefully, and remove as little metal as possible.

Once you're satisfied that the center table is flat, lay your straightedge across the joints between the center table and both extension wings. To flatten these two long joints, slightly loosen the mounting bolts that hold the extension wings in place, then tap the extension wings up or down with a soft-faced mallet or a hammer and wood block until the surfaces are flush. Retighten the bolts and check the joints again with a square.

Using winding sticks. Check the center cast-iron portion of the saw table for warp with a pair of hardwood winding sticks. Set the sticks along the edges of the saw table and inspect the table across its width (A), then across its depth (B).

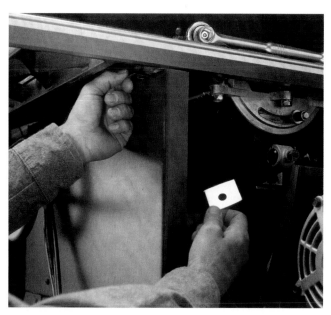

Tabletop warp less than 1/16" across the surface may be correctable by inserting thin metal shims underneath one or more corners of the saw table, where it bolts to the saw base. Your odds for correcting warp in this way improve if the saw table is made of softer, more flexible aluminum alloys. For more severe warp, replace the saw table or have it reground at a machine shop.

Check the entire saw table for flatness by laying the straightedge across the full tabletop. Check to see that the extension wings don't bow up or down from the center table. If they do, and your extension wings are made of pressed steel, loosen the bolts that fasten them to the fence rails, pull the extension wings up or down to flatten them, and retighten the bolts. If the extension wings are made of cast iron or another solid metal alloy, you'll have to use thin plastic or metal shims in the joints to tip the wings up or down. Bolt the wings tightly to the center table again with the shims in place. Trim the shims flush.

Drag a long straightedge over the center cast iron portion of the saw table to identify high and low areas. Then check the joints between the center table and the extension wings. If the joints are uneven, loosen the extension wing bolts and tap the joints flush with a rubber mallet.

Circle high areas on the saw table, then carefully grind them flat with a fine-grit sharpening stone lubricated with water or fine-grade machine oil. Try to remove as little metal as possible—your goal is to just even out these surface irregularities.

Shimming extension wings. If your saw has cast iron extension wings and they tilt up or down, flatten them to the saw table by inserting plastic shims into the joint between the extension wing and the center table.

Adjust the Throat Plate

The last step in flattening the saw table is to adjust the removable throat plate. Most saw throat plates have four Allen screws that adjust the throat plate up and down to sit flush with the table surface. On some saws, these setscrews may actually fasten the throat plate to the saw table.

Regardless of which type of throat plate your saw has, a flush fit with the top of the table is important. A flush throat plate won't snag workpieces as they are fed into the blade, and it won't allow wood to catch on the back edge of the throat plate as workpieces exit the blade. Check to see if your throat plate needs adjustment by laying a short straightedge across the throat plate and shining a flashlight along the edge where the throat plate meets the straightedge. Adjust the throat plate setscrews until the sliver of light disappears. Be sure the throat plate is flush with the saw table both widthwise and lengthwise.

Check the throat plate for a flush fit with the saw table by shining a light behind a steel rule laid over the throat plate. Any sliver of light that appears beneath the rule indicates that the throat plate is set too low. To adjust, turn the setscrews. Light that shows between the rule and the table means the throat plate is too high.

How to Make a Zero-Clearance Throat Plate

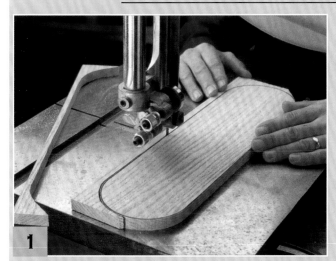

1

Plane a length of hardwood to a thickness that matches the depth of the throat plate recess in your saw table. Use the factory throat plate as a template for tracing the throat plate shape onto your hardwood blank. Trim the blank to shape with a band saw or scroll saw, and sand the cut edges smooth.

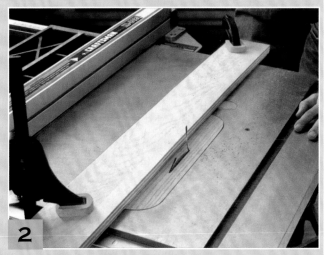

2

With the saw blade lowered, set the new throat plate blank into the saw table recess. Clamp a piece of scrap wood on top of the throat plate to hold it in place. Turn on the saw and crank the blade up slowly until it cuts through the throat plate. Raise the blade to full height.

Option: Replace, Rather Than Adjust, Your Throat Plate

The trouble with most standard throat plates is that the slot for the blade is too wide. Wide gaps around the blade will cause the blade to splinter some woods (called *tearout*) because the wood isn't supported next to the blade. Rather than shearing the wood cleanly, the blade begins to tear the wood fibers as it cuts.

Another problem with wide blade slots is that narrow scraps of wood can actually get lodged in the slot between the throat plate and the spinning saw blade. If a substantial scrap should catch on the blade teeth, the blade could throw the cutoff scrap back toward you. A good solution to these problems is to install a shop-built throat plate with a blade opening that's only slightly wider than the thickness of your

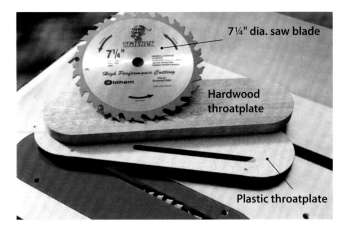

7¼" dia. saw blade

Hardwood throatplate

Plastic throatplate

Build your own zero-clearance throat plates from scrap hardwood, or buy a preformed plastic throat plate blank from a woodworking supply house. Aftermarket plastic throat plates come shaped to fit specific saws, and some, like the one shown here, have a slot cut partway through to make room for the blade. You may need to install a smaller 7¼"-diameter circular blade on your saw in order to cut the initial kerf opening, if your 10"-diameter blade doesn't drop below the saw table far enough to clear the bottom of the zero-clearance throat plate. Case in point: the blade installed in the saw shown here is lowered as far as it will go, and it barely clears the bottom of the red throat plate.

blade—commonly called a *zero-clearance throat plate.* They're easy to make. Here's how:

Select a scrap of hardwood that's larger than the throat plate opening in your saw table and thicker than the throat plate recess in the saw table. You can also purchase plastic throat plate blanks from woodworking supply catalogs that are shaped to fit a variety of different saws (see photo, above and page 14). If you choose to make your own hardwood throat plate, plane the wooden blank so it matches the exact depth of the throat plate recess in the saw table.

Use the original throat plate that came with your saw as a template to draw the throat plate shape onto your hardwood blank. Cut out the throat plate shape with a band saw or scroll saw and smooth the edges with a sander.

If the throat plate on your saw fastens to the saw table with screws, drill countersunk pilot holes in the wood throat plate for the screws so the screw heads will sit below the surface of the throat plate once it is installed on the saw.

Since the goal of a zero-clearance throat plate is to match the throat plate opening to the blade thickness, use the saw blade to cut its own blade slot in the throat plate. First, raise the blade to full height and square it to the saw table. Then crank the blade all the way below the saw table and set the zero-clearance throat plate in place.

If the blade won't clear the throat plate from below, which is commonly the case, you'll have to install a smaller-diameter blade in the saw to start the kerf cut. On most saws, a 7¼" circular saw blade with a ⅛" kerf will do the trick, but be sure the thickness of the blade matches the kerf of the larger blade you plan to use with the throat plate. Hold down the throat plate by positioning a board over the throat plate and clamping the board to the saw table. Start the saw and raise the blade slowly until it cuts through the surface of the throat plate. Lower the blade again, unplug the saw, and change to your larger diameter saw blade. Finish cutting the kerf opening in the throat plate by raising the blade gradually through the throat plate to its full height.

Don't Toss Your Old Throat Plate

Here are a couple things to keep in mind when using zero-clearance throat plates. First, you won't be able to tilt the blade with a zero-clearance throat plate in place. The narrow kerf will obstruct the blade as it tilts on the trunnions, so switch back to your original throat plate when making bevel cuts. Second, you may need to make several throat plates if you routinely use different blades—blade kerf widths may vary by blade type and manufacturer.

Single-Use Throat Plates

A quick alternative to building zero-clearance throat plates is to simply clamp a piece of tempered hardboard or ¼"- to ½"-thick plywood to the saw table over the throat plate area. Raise the blade until it cuts through the scrap board. Be sure to use a piece long enough to extend across the depth of the saw table and wide enough to support the workpieces you are cutting. Keep your hands clear as you cut the kerf.

CHECK BLADE AND MITER SLOT ALIGNMENT

For all-around accurate cutting, it's not enough that the rip fence and blade are parallel with one another. The blade must also be parallel to the miter slots. If it isn't, the blade will make skewed crosscuts when you are using the miter gauge, because it won't be cutting in a plane that's perpendicular to the miter gauge. When the blade isn't parallel to the miter slots, the condition is called *blade heel* (see Understanding Blade Heel, page 44).

Although blade heel is a common problem on new saws that have never been tuned, it's not something you can see without careful inspection. If the blade is off by even ¹⁄₃₂" from front to back, it will affect how the saw cuts. Once you've corrected for blade heel, you'll probably never need to adjust for heel again unless the saw is disassembled.

When tuning your table saw, use the miter slots as your standard of reference for aligning both the blade and the rip fence. This way, not only will the blade and rip fence be parallel with the miter slots, they also will be parallel with one another.

Tip

A 2' straightedge will work for most saws, but it is better to use one that stretches across the full depth of your saw table. It's important that the straightedge be absolutely flat along its edge, because it will establish the exact geometric plane of the blade relative to the miter slots. You'll use the straightedge to measure and compare the distances between the blade and a miter slot on both the infeed (front) and outfeed (back) ends of the saw table. The farther away from the blade you take your measurements, the more exaggerated the differences in measurements will be, so you'll be able to detect even the slightest heeling.

1

To test for blade heel, unplug the saw, then raise the blade to full height. Set a try square on the saw table and against the saw blade body. Be sure the square isn't touching the carbide blade teeth; the square must sit flat against the blade body. Crank the blade-tilting handwheel one way or the other until the blade is square to the table. If the handwheel stops before you reach the point where the blade is square, see "Setting Blade Tilt Stops," page 46.

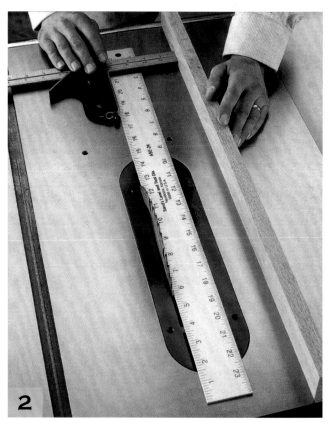

2

Lay a long metal straightedge against the blade body to mark its rotational plane. Set a piece of scrap wood that is about 1½" wide x 3' long into the miter slot (it should fit snugly). Set a combination square against the straightedge near the infeed edge of the saw table. Slide the rule until it touches the scrap. Take a reading at the front and back of the saw table. The measurements should match; if not, the blade is heeling. To correct the blade heel, see pages 44–45.

Understanding Blade Heel

When the blade isn't parallel with the miter slots, the condition is called *heel*. A blade that heels will cut at an angle to both the miter gauge and the rip fence, so both rip cuts and crosscuts will be inaccurate. Blade heel is hazardous in ripping situations when the outfeed side of the blade heels toward the rip fence. In this instance, every time you make a rip cut, you are essentially forcing your workpiece into a funnel, which binds the workpiece between the fence and the blade, increasing your chances for kickback. The trouble with heeling on the infeed side is that the blade will pull the workpiece away from the fence. In either case of heel, the blade actually cuts twice—on both the upstroke and the downstroke. Cutting twice heats up the blade, resulting in burned edges on your work.

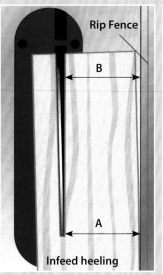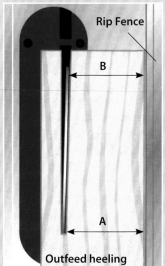

Infeed heeling occurs when distance A is shorter than distance B. The blade will tend to pull workpieces away from the rip fence.

Outfeed heeling forces workpieces into a narrower opening at the back of the blade. Distance B is shorter than distance A.

CORRECTING BLADE HEEL

To correct infeed or outfeed blade heel, you'll need to shift the position of the arbor relative to the saw table. The process for making this adjustment varies, depending upon how the arbor mounts to the saw. On most cabinet saws, the arbor assembly bolts to the saw cabinet, not the saw table. To shift the arbor, locate the bolts that hold the saw table to the cabinet and loosen all but one slightly. (Usually there are four bolts.) Then tap the saw table until the blade is parallel to the miter slots. You may need to loosen the fourth bolt a touch to make pivoting easier.

On contractor's and direct-drive jobsite saws, as well as some smaller cabinet saws, the arbor assembly bolts to the saw table, not the saw base. You'll need to loosen three of the four bolts that secure the trunnions to the underside of the saw table. Tap the cradle assembly from behind one way or the other with a hammer and wood block to shift the arbor and blade into alignment with the miter slots. Remember, you may be able to gain better access to the cradle assembly by reaching through the throat plate opening. Regardless of the saw type, work carefully and check your progress against the combination square and rule. Once you find the correct arbor position, retighten the bolts evenly.

A good dynamic test to check your blade alignment is to simply listen for signs of heeling. To do this, first be sure the bar on your miter gauge fits snugly in the miter slot (see

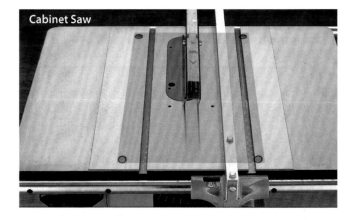

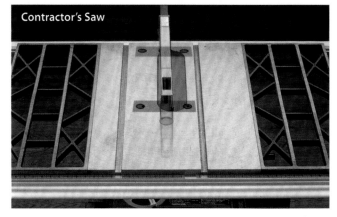

The shaded areas and dark circles on these saws represent the locations of the bolts you'll need to loosen in order to correct a heeling blade. Once the bolts are loosened on a cabinet saw, shift the whole saw table to realign the miter slots with the blade. On a contractor's saw, loosen the bolts and shift the trunnions instead.

How to Adjust Trunnions

Once you've loosened the trunnion bolts on a contractor's saw, use a hammer and wood block to move the entire trunnion assembly to the right or left. Doing so will alter the blade's relationship to the miter slots, which is necessary if the blade is heeling. Rap lightly and check your progress often.

Option: Depending upon the make of your saw, you may be able to gain better access to the trunnion assembly by working through the throat plate opening in the saw table.

page 49). Remove the saw's drive belt(s). Clamp a length of scrap wood to the miter gauge fence so the blade teeth just graze the scrap when you spin the blade by hand. Raise the blade to full height. Slide the miter gauge to the infeed side of the blade and spin the blade hard. Listen to the sound the blade makes as it rubs against the scrap. If the sound is even all around, you've corrected blade heel. If the pitch changes, determine the area of the blade where the change occurs and mark a blade tooth in this area with chalk or a felt-tipped marker.

Slide the miter gauge and scrap to the outfeed side of the blade and spin the blade again. If the blade area you marked grazes the scrap and makes the same noise as it did when you tested on the infeed side, you've corrected the heeling condition, but the blade is warped in the area you've marked. If the sound is louder or softer in the marked area than it was when you tested on the infeed side, the blade is still heeling.

Continue your adjustments until the blade no longer heels. If it seems your blade is warped, check the amount of warpage (called *runout*) with a dial indicator (see pages 52-53).

Check your heeling adjustments by spinning the blade and allowing it to rub against a wood block clamped to the miter gauge. Set the block near the front of the blade, spin the blade, and listen. Then slide the miter gauge and scrap to the back of the blade and spin. If the pitch you hear is consistent all around, the blade is no longer heeling. If the rubbing sound gets louder or softer, mark the area on the blade where the pitch changes. Changes in pitch likely indicate that the blade is still heeling.

The location of blade tilt stops, which help set 45° and 90° blade tilt settings, varies from saw to saw. On this contractor's saw, one stop is mounted on a stabilizer bar on the cradle assembly, and the other stop is situated at the end of the front trunnion. Check your owner's manual to locate the stops on your saw.

Table-mounted tilt stops. Some manufacturers make the process of setting blade tilt easier by mounting the stop screws in the saw table rather than underneath. To adjust blade tilt, align the blade to a drafting triangle or try square and turn the stop screws until you feel resistance.

SETTING BLADE TILT STOPS

Most table saw cuts take place with the blade set at either 90° or 45° to the saw table. To help speed up the process of setting these angles, better-quality saws are equipped with adjustment screws or bolts, called tilt stops, that limit the path of the trunnions as they tilt. When the tilt stops are correctly set, you should feel resistance in the handwheel when the pointer on the blade tilt scale reaches the 90° and 45° marks on the scale. If you can't reach either blade angle before the handwheel stops turning, or if you can crank past these marks, one or both tilt stops need adjustment.

To set the 90° stop, crank the blade to full height, lay a try square against the blade body, and adjust for square. Check your owner's manual to locate the 90° tilt stop (usually located on the end of the front trunnion). The stop may be a simple setscrew or a bolt and nut. Turn the stop in or out until the handwheel stops the blade at 90°. Set the 45° blade stop in the same fashion. This stop may be located near the 90° stop on the trunnion or on the opposite end of the trunnion altogether. Check your owner's manual to be sure. Use a 45° drafting square or a speed square to set the blade to precisely 45° on the saw table before you adjust the stop.

A few saw manufacturers simplify the job of adjusting the blade stops by recessing Allen screws into the saw table near the throat plate. The Allen screw system replaces the standard stops located on the trunnions, so you can adjust the blade without reaching under the saw table. One Allen screw sets the 90° blade stop and the other screw sets the 45° stop. To adjust the blade tilt, crank the blade to either 90° or 45°, check the blade setting with a square or triangle, and turn the Allen screw until you feel resistance.

Once you've established accurate 90° and 45° blade stops, reset the blade tilt pointer on the front of the saw so the scale reads accurately when the blade is tilted to each stop. You may have to remove a handwheel to adjust the pointer. As a rule, use the blade tilt scale to "ballpark" your blade angle settings. Don't rely on the blade tilt scale for setting precise blade angles. Where accuracy is important, always set and check your blade angle against a truing instrument—a bevel gauge set to the angle you want, a drafting triangle, or a square—before you make your cut. Then cut a test piece first to verify your accuracy.

Dynamic Test for Checking 90° Tilt Setting

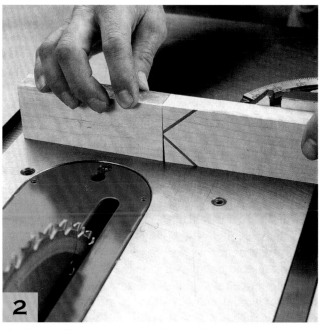

Once you've adjusted your 90° blade tilt stop, check the setting for accuracy by crosscutting a piece of scrap using the miter gauge as a guide. Mark the face of the scrap first to keep the orientation of the parts clear once you cut it in two.

Flip the cutoff piece over and set the cut edges together on the saw table. If the tilt stop is set accurately, the cut edges should meet flush. If a gap appears, it is double the amount the blade is off of 90°. Adjust your 90° tilt stop and test again.

Dynamic Test for Checking 45° Tilt Setting

Conduct a similar dynamic test as the one outlined above to check your 45° blade tilt setting. Tilt the blade until it stops at 45°. Mark one face of a piece of scrap and crosscut it with the workpiece held against the miter gauge.

Flip one of the cut edges and set the scrap pieces together to form a mitered corner. If the 45° blade tilt stop is set accurately, the scrap pieces should form a 90° corner. If they don't, adjust the tilt stop and try the dynamic test again.

ALIGN THE RIP FENCE AND MITER SLOTS

Once you've adjusted the blade so it's parallel to the miter slots and square to the table, do the same for the rip fence. Set a pair of 1½"-wide wood blocks into a miter slot, one near the front and one near the rear of the saw table. The blocks should fit snugly in the miter slot. Slide the rip fence until it rests against one or both of the blocks. If the fence is parallel to the miter slot, it should touch both blocks. Otherwise, the fence needs to be adjusted.

Depending upon the style, adjusting your saw's rip fence involves loosening two bolts located on either the top or the outside face of the fence body. Refer to your owner's manual to locate these adjustment bolts. Loosen them and shift the fence body to the left or right until the fence touches both wood blocks at the same time. Then retighten the bolts and check the alignment again. If your fence works properly (and some may not), you should be able to slide the fence anywhere along the fence rails and lock it down parallel to the blade. Be aware that not all fences are created equal. Some fences are chronically difficult to keep tuned. If your fence is one of these, you may want to consider buying a replacement fence. For more on replacement fences, see pages 138–139.

Since the rip fence gets moved frequently and even removed, it's a good idea to spot-check your fence alignment from time to time. Save the miter slot wood blocks for future testing.

To align your rip fence to a miter slot, slip a couple of scrap blocks into a miter slot. The block should fit snugly. Slide the rip fence up to the blocks. If the fence doesn't touch both blocks, loosen the adjustment bolts on the rip fence, tap the fence body until it touches the blocks, and tighten the bolts.

SQUARE THE RIP FENCE TO THE SAW TABLE

When a rip fence is locked down to the table, the face of the fence should be perpendicular to the saw table. Otherwise, workpieces won't sit flat against the fence and square with the blade. Test how close to perpendicular your fence is to the saw table using a try square. Some better-quality fences will have setscrews to help you make this adjustment (see left photo under "How to Square a Rip Fence"), but most fences will not. If your fence has no setscrews and it isn't square to the table, you may have to attach an auxiliary wood fence to the rip fence and shim it to square (see right photo under "How to Square a Rip Fence").

How to Square a Rip Fence

Some rip fences have setscrews that allow you to square the fence to the saw table. On these fences, set a square against the fence and turn the screws in or out until the fence rests flush against the square.

If your rip fence has no adjustment screws for squaring it to the table, attach an auxiliary wood fence to the rip fence, and insert paper shims until the wood fence is square to the saw table.

TUNING THE MITER GAUGE

A miter gauge is essentially a sliding crosscut fence for your saw. To be precise, the miter gauge bar needs to slide along the miter slots smoothly and without side-to-side play. You can also adjust the protractor head for accuracy at both 90° and 45° if your gauge comes with stop screws at these angle settings.

To test the miter gauge bar, set it in a miter slot and try to wiggle it from side to side. It's common for the bar to be slightly undersized for the miter slot. The miter gauge bar will loosen in the slot as it wears from sliding back and forth in the miter slot. You can improve the fit of the miter bar easily by tapping pairs of dimples along either bar edge with a hammer and center punch. Space the dimples every inch or so along the full length of the bar. Try the fit of the bar in the slot; if it doesn't fit in the miter slot or drags when you slide it back and forth, knock down the ridges of the dimples slightly with a fine metal file.

Tightening Loose Miter Bars

Take the slop out of your miter gauge's slide by punching a series of dimples along the edges of the miter bar with a metal punch and a ball peen hammer. If the bar fits too tightly after you do this, file the tops of the dimples down slightly. The bar should slide smoothly without side play.

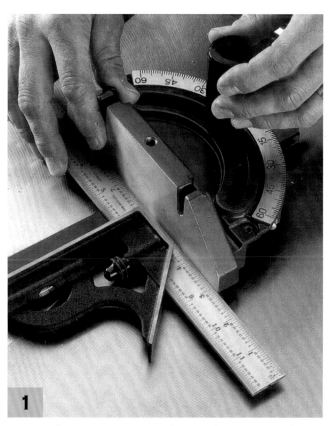

1

Use a combination square as a guide to set the miter gauge fence so it is 90° to the metal bar that slides in the miter slots. Rest the head of the square against the bar and the rule against the fence.

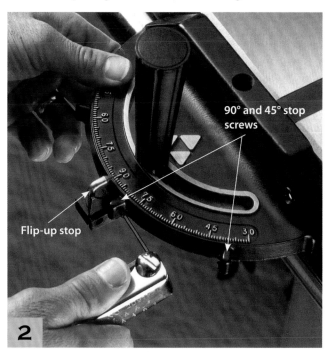

90° and 45° stop screws

Flip-up stop

2

Most miter gauges have stop screws that allow you to set the fence to 90° or 45° quickly. Tip the flip stop up and turn the 90° stop screw until it touches the flip stop. Then pivot the miter fence against a speed square or plastic drafting triangle, set the fence for 45°, and adjust the 45° stop screw against the flip stop as well.

TUNE THE SPLITTER OR RIVING KNIFE AND PAWLS

Splitters and riving knives need to line up perfectly behind the blade, or they'll hang up workpieces as wood exits the blade. Those that come as standard table saw equipment are made of a piece of sheet metal, which can be adjusted to the left or right behind the blade: here's how. Loosen the nuts that hold the splitter/riving knife assembly to the back of the saw. (Some better-quality saws will also have an adjustable mounting plate beneath the throat plate on the cradle assembly that holds the front of the splitter or riving knife in place. You may need to loosen the bolts on this mounting plate as well.) Set two straightedges against the blade body (not against the carbide teeth) on either side. Shift the splitter or riving knife until it sits between, but does not touch, either straightedge. Retighten the mounting bolts.

Test the effectiveness of your anti-kickback pawls by making a rip cut down the center of a piece of ¾"-thick scrap about 3' long. Stop the cut and shut off the saw when about 1' of the workpiece has passed through the blade. Unplug the saw and lower the blade. Grab ahold of the workpiece on the infeed side of the blade and try to pull it back toward yourself. If you can pull the workpiece backward out of the cut, the anti-kickback pawls need tuning. By design, the pawls should drag along the top of the workpiece as it passes through the blade from front to back but dig into the wood if kickback occurs.

The bottom edges of the anti-kickback pawls should sit as close to the saw table as possible when not in use. The best scenario is that the tips of the pawl teeth actually touch the saw table. On some saws with older splitter assemblies, the pawls may float ¼" or more off the table, which renders them useless for protecting you from kicked-back stock that is less than ¼" thick. You may need to modify how the splitter mounts to the saw so the pawls will rest closer to or against the saw table. Elongating the splitter bolt holes may do the trick, in order for the entire splitter to sit lower on the saw table.

If the pawls touch the table but they still allow you to slide workpieces back out of the cut, the problem may be that the points on the pawls are too dull to "bite" into the wood. To remedy the problem, remove the pawls and sharpen the points with a flat file, then reinstall the pawls on the splitter or riving knife assembly. The pawls may also be outfitted with springs that are too weak to hold the pawls firmly against the workpiece. If this is the case, install stiffer springs.

You may be able to replace the splitter and anti-kickback pawls on your older saw with a better-quality, removable splitter, especially if you own a popular-model contractor's or cabinet saw. For more on aftermarket splitters, see page 141.

Anti-kickback pawls removed for clarity

To adjust a splitter or riving knife, sandwich the saw blade between two straightedges. If the splitter or riving knife is out of alignment, loosen the bolts that secure it to the saw and shift it until it touches neither of the straightedges. Retighten the bolts.

Tuning anti-kickback pawls. If the anti-kickback pawls mounted on your older saw's splitter fail to grab your workpieces, you may be able to improve pawl performance by sharpening the teeth with a file. If this doesn't help, set the entire splitter lower on the saw table, or buy stiffer springs to hold the pawls in place.

REDUCING SAW VIBRATION

A saw that vibrates excessively will produce rougher cuts than one that runs smoothly. This is because vibration makes the blade flex back and forth as it spins on the arbor, so the blade makes wider, uneven cuts. Over time, vibration can also loosen fasteners and other parts on the saw, but the primary problem with table saw vibration is how it affects cutting performance. If you notice your saw vibrating more than normal, the two likely culprits to check are the belts and pulleys or the arbor and blade.

Tuning Up Motor and Arbor Drive Pulleys

On contractor's and cabinet saws, the pulleys on the motor and arbor must line up with one another. Otherwise, the drive belt or belts will ride unevenly in the pulleys and set up vibration in the arbor assembly, which will transfer right into the blade. Also, the motor shaft and the arbor shaft must be parallel to one another to provide even tension on the belts.

Check pulley alignment by laying a straightedge against both the outer rim of the pulley on the arbor and the counterpart rim on the motor shaft. The straightedge should be flat against both rims. If it isn't, loosen the setscrew on the motor shaft pulley and slide the pulley in or out along the shaft until all the pulley rims line up. If you still can't get the pulley rims to align, the arbor and motor shafts are not parallel with one another. Adjust the shafts by loosening the bolts that hold the motor on the motor mounting plate, and shift the motor slightly until the pulleys come into alignment.

Check the condition of the drive belt or belts on your saw while you are aligning the pulleys. The tapered inner surfaces of the belts should be smooth and free of splits or missing chunks of rubber. The belts should ride evenly on the pulleys, just below the top edges of the rims. If your belts look dried out, worn, or damaged, or if they slap and vibrate when the saw is running, replace all of them at one time with the correct factory replacements recommended for your saw. Don't make the mistake of substituting automotive fan belts—they're engineered for different tolerances, speeds, and pulley geometries.

Adjusting Drive Pulleys

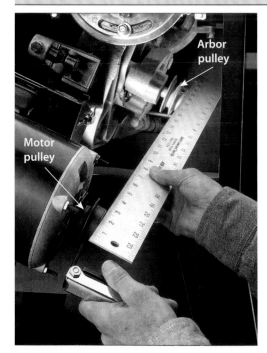

Shift the motor pulley. Set a straightedge against the rims of the motor and arbor pulleys to check their alignment. If the straightedge won't sit flat against both pulleys at the same time, loosen the screw that holds the motor pulley in place and slide it in or out on the motor shaft until the pulleys line up.

Shift the whole motor. If adjusting the motor pulley still doesn't line up the pulleys, loosen the motor mounting bolts and shift the motor to make up the difference. It may help to loosen up only three of the four mounting bolts, and use the fourth bolt as a pivot point to keep the motor from moving too far.

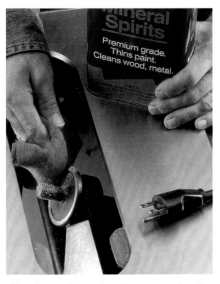

Check and replace drive belts. Dried out, worn, or cracked drive belts contribute to saw vibration. Replace a worn drive belt with a matching belt designed specifically for your table saw.

Stabilize the rear trunnion. You may be able to reduce vibration emanating from the arbor and trunnions by fastening a C-clamp to the rear trunnion and the back of the cradle assembly.

Cleaning a dirty arbor flange may be all it takes to eliminate excess vibration. The blade must sit flat against the rim of the arbor flange, or it will wobble and lead to more vibration. Use mineral spirits and a soft cloth to remove accumulated wood pitch or sawdust.

Correct Arbor and Blade Runout

If you've checked other possible sources of vibration and the saw still vibrates excessively, turn your attention to the blade and arbor. The arbor and blade must spin in the same flat plane for the blade to cut accurately. If either the blade or the arbor wobble out of the flat plane of rotation, the saw will vibrate. This wobbling is called *runout*.

Remove the saw blade from the arbor and inspect the arbor flange closely (see far right photo, above). It only takes a small amount of dirt, compacted sawdust, or a burr on the metal to prevent the blade from butting tightly against the flange. Clean the flange with mineral spirits and a bristle brush first. Then slide your finger around the flange, feeling for imperfections. You are most likely to find nicks or burrs on the arbor when it is new. File off any nicks or burrs.

While you inspect the flange, check the arbor for worn bearings, which can also lead to arbor runout. Remove the drive belt and spin the arbor. If you hear clicking or grinding, one or more arbor bearings are bad and should be replaced.

Reinstall the blade and see if the vibration disappears. If cleaning the arbor flange doesn't solve the problem, turn your attention to possible blade runout.

If you've just installed a new or re-sharpened blade on your saw and notice more vibration, the problem is likely coming from the blade. The blade's arbor hole may be slightly off center or larger than the diameter of the arbor shaft, which will make the blade spin out of round on the arbor shaft. It's possible that if the blade was just sharpened or subjected to a trauma like overheating or kickback, the blade body may have fAllen out of balance. Try a different blade in the saw to see if another blade remedies the problem. Shifting the blade to a new location on the arbor shaft may also improve runout, so loosen the arbor nut, rotate the blade a quarter turn on the arbor, and retighten the nut. Or you could add a pair of blade stiffeners (see bottom photo, page 53) to the blade, which sandwich the blade on one or both sides and stiffen the blade body.

Tip

Choose a fine-toothed flat file and remove as little metal as possible, to keep the flange face flat. You can also rub away irregularities with a fine-grit sharpening stone.

How to Test for Arbor and Blade Runout with a Dial Indicator

Magnetic base

Arbor flange

After you've cleaned the arbor flange, removed any burrs and tried several different blades in an attempt to reduce vibration, persistent vibration indicates that either the arbor or the blade has excessive runout. Use a dial indicator to measure the amount of runout on your arbor flange first, then test your blades. A dial indicator is a machinist's tool that measures the amount of deflection on a shaft or blade in thousandths of an inch.

To test for arbor flange runout, pick up a dial indicator that clamps to a magnetic base. Remove the saw blade. Set the dial indicator base on the saw table close to the blade. Adjust the apparatus so the contact point on the dial touches the flat face of the arbor flange. Set the dial to zero. Then rotate the arbor shaft slowly by hand as you watch the dial. The overall amount of deflection from zero should not exceed .003. If it does, consider having the arbor and flange re-trued by a machine shop or replacing the arbor altogether. Any more deflection than this will translate to rougher cuts on your workpieces.

To test blade runout, mount the blade to the arbor and tighten the arbor nut. Set the dial indicator in position with the contact point against the blade as far out to the edge of the blade as possible. Keep the contact point far enough inset so it doesn't accidentally catch in the blade tooth gullets (the "valleys" between the teeth). Then rotate the blade slowly. All blades will have some runout, and runout will slowly increase as the blade heats and cools repeatedly. Blade runout should not exceed .012 on 10" blades. If your blade does, it's time to buy a new blade.

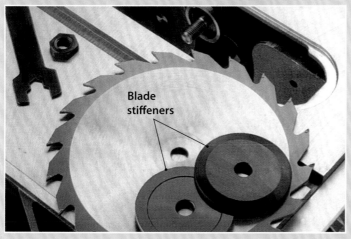

Blade stiffeners

Minor blade runout can be improved by installing metal blade stiffeners on each side of the blade. Stiffeners are sold in different diameters. Keep in mind that larger diameter blade stiffeners will make contact with workpieces if the blade is raised too far.

PERIODIC CLEANING AND LUBRICATION

A little routine maintenance goes a long way toward prolonging the life of your saw. In addition to regular tune-ups, your saw will benefit from regular cleaning and lubrication. Here are a few tips to follow.

Vacuum the Saw Base, Internals, and Motor

Table saws create a tremendous amount of sawdust, and it will accumulate quickly unless you clean it up every time you are finished sawing. Sweep off the tabletop and wipe off the fence rails as you work to keep the rip fence and miter gauge sliding smoothly. After you finish cutting for the day, take time to brush out and vacuum inside the saw base (see photo, right), particularly around the motor's cooling vents and along the trunnions and arbor assembly where the worm gears engage.

Clean and Polish the Saw Table

Treat the metal tabletop (including the miter slots), fence rails, and miter gauge with a coat of paste wax every now and then. Apply wax frequently if your shop is in a damp basement or garage where the level of humidity is high. Waxing these parts will not only keep the bare metal from rusting but also will keep workpieces and saw parts sliding smoothly during saw use. Spread the wax evenly, rub it in, and allow it to dry to a haze, then buff off the excess with a clean towel. Choose a paste-style wax formulated for automobiles or furniture, but check the label to be sure the wax does not contain silicone. Silicone on the saw table will absorb into wood as you cut it. Once absorbed into your workpieces, silicone will repel wood finishes like stain and varnish later on.

Use a synthetic pad and naval jelly or automotive rubbing compound to remove minor rust and pitting when it occurs and mineral spirits to remove wood pitch. Follow these treatments immediately with a coat of wax.

Lubricate Worm Gears and Moving Parts

Eventually you'll have to remove gummed-up, dried sawdust and pitch deposits from the trunnions and worm gears, as well as other nooks and crannies inside the saw base and behind the blade. Resinous woods like pine and woods that aren't thoroughly dry will gunk up a saw even faster. Use a soft brass- or plastic-bristle brush and mineral spirits to soften and scrape away impacted sawdust. Once the parts are clean and dry, lubricate the worm and trunnion gears and other pivot points with powdered graphite, a non-silicone spray lubricant, or a dab of paste wax. If you use graphite, keep it away from the saw table, or it will act like silicone on wood and repel wood finishes. Although you may be tempted to use it, steer clear of petroleum greases to lubricate your table saw. Grease will only attract and trap the grit and sawdust.

Lubricate both sets of worm gears

Clean and Protect Blades from Rust

It's important to keep saw blades clean and rust-free. When pitch and gum deposits accumulate on the blade, it heats up more quickly and is more likely to leave burn marks as it cuts. Dirty, gummed-up blade teeth also attract more pitch and gum deposits, so the problem compounds over time. And some wood resins will eventually corrode carbide teeth. Accumulated wood pitch also affects the blade's balance as it spins.

Commercial blade cleaners do a fine job of removing blade deposits. However, you can save money by simply soaking the blade in spray-on oven cleaner for about an hour, then scrubbing the blade with a synthetic abrasive pad for pots and pans and wiping it clean with a towel. Once the blade is clean, inspect it carefully under a bright light for cracks. Use a magnifying glass to examine the carbide teeth for chips, dulled edges, or cracks in the brazing that secure the teeth on the blade. Then protect the blade from rust with a light coating of paste wax or oil.

Table Saw History

Double-Blade Flooring Saw

Carroll's Patent Gang Saw, manufactured in the early 1870s, was designed to rip-cut 4"- to 6"-wide planks for flooring. It featured two ripping blades on the arbor—the right blade fixed in place and the left blade adjustable to the right or left. By outfitting multiple blades on the arbor, the saw could rip up to three widths of flooring simultaneously from a single board—one plank between the rip fence and right blade, the second between the two blades, and the third to the left of the left blade. A handle mounted on the front of the saw made adjustments to the left blade possible without wrenches, to economize setup time.

This belt-driven saw was outfitted with a feed roller behind the blades to pull lumber through the blades. A published review of this machine estimated that it could saw between 10,000 and 12,000 board feet of flooring in a ten-hour workday, double the productivity of other flooring saws at the time.

Carroll's Patent Gang Saw (circa 1873)

CHOOSING
SAW BLADES

Tuning up your saw will dramatically improve the accuracy of your cuts, but don't overlook the importance of choosing the right blade for the job. At first glance, blades look basically the same, but that doesn't mean one blade will perform all cuts equally well. Differences in the shape and number of blade teeth distinguish blades meant for ripping, crosscutting, trimming laminates, or cutting plywood. You won't need a drawer full of saw blades to be properly outfitted, but choosing a few blades carefully and using the right blade for the task will go a long way toward producing cuts that require little sanding or planing.

To make wise blade choices, you need to know a few details about blade anatomy and types of blades, as well as how to tell a quality blade from a poor one. The next few pages should provide all the information you'll need to buy a hardworking, comprehensive set of blades for your table saw.

KNOW THE PARTS

A blade begins as a plate of steel that is stamped or laser-cut to shape, forming the *blade body*. The steel is heat-treated to temper it to a certain hardness, then tensioned. *Tensioning* is a process of compressing a ring around the circumference of the blade body near the *rim* to help the blade dissipate stress and heat more efficiently so it doesn't warp. An *arbor hole* (usually undersized) is punched into the center of the blade body, then reamed wider to a precise diameter. The blade body is ground flat and, on most blades, uniformly thick. (Some blades with steel teeth, called *planer blades*, are "hollow-ground," so they are thicker near the center and ground thinner out to the teeth.) The *gullets* and *shoulders* are also ground smooth.

Located between the teeth, gullets serve a cooling function, like the thin fins on a radiator. But their main job is to evacuate chips and sawdust from the saw kerf. Shoulders reinforce the area behind each tooth. Some blades are designed with special *anti-kickback tips* on the shoulders, to limit how deeply each tooth bites into the wood.

Anatomy of a Blade

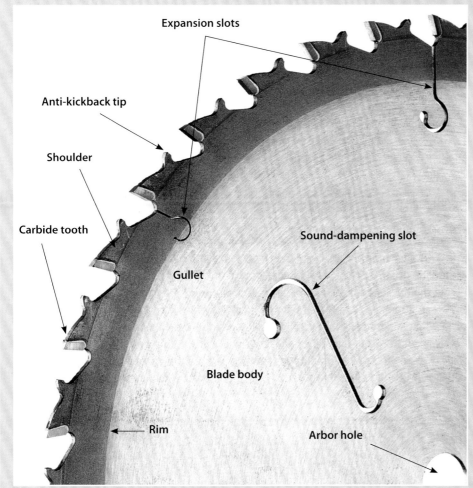

Expansion slots

Anti-kickback tip

Shoulder

Carbide tooth

Sound-dampening slot

Gullet

Blade body

Rim

Arbor hole

Saw blades are made in a variety of styles and tooth counts, each engineered for specific cutting tasks. Most saw blades feature carbide teeth, which hold an edge longer than steel teeth. Better-quality blades will have a number of expansion slots cut into the blade body to keep it from warping as it heats up. Some manufacturers also add sound-dampening slots to diffuse vibration that would create excessive blade noise.

Understanding Hook Angles

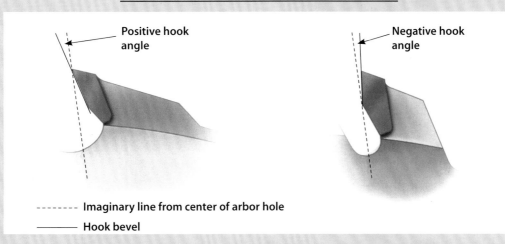

Positive hook angle

Negative hook angle

- - - - - - Imaginary line from center of arbor hole

——— Hook bevel

Hook angle is a measure of how far forward or backward carbide teeth tip on the blade shoulders. Blades have "positive" hook angles when the teeth tip forward and "negative" hook angles when teeth tip backward. The degree of hook is measured from an imaginary line (considered 0°) extending from the center of the blade arbor hole to the tip of the carbide tooth. Hook angles range from 20° (ripping blades) to -6° (laminate-cutting blades). Teeth with positive hook angles cut more quickly and produce larger chips than teeth with negative hook angles.

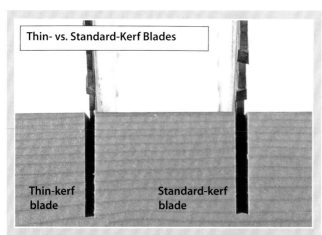

Thin- vs. Standard-Kerf Blades

Thin-kerf blade

Standard-kerf blade

The usual kerf width for standard saw blades is about ⅛" to ³⁄₁₆", but the thickness of a thin-kerf blade is closer to ³⁄₃₂", so it removes less material when it cuts. The big advantage to using a thin-kerf blade is that you'll need less saw power to perform the same cut as a standard-kerf blade. Consider buying a thin-kerf blade if you own a saw with a motor less than 1½ h.p. The downside to thin-kerf blades is that they are more likely to deflect when cutting thick or dense woods.

Carbide Teeth

Carbide teeth are incredibly hard, but they will dull and even chip as the teeth wear. Both of the blades shown here have alternate top bevel teeth, but notice the corroded, chipped, and missing tips on the top blade compared with the new blade on the bottom. Carbide teeth can be resharpened several times on better blades, but the carbide teeth on the top blade should be replaced. Have your carbide-tipped blades serviced by a qualified saw sharpener, or send the blades back to the manufacturer for repairs.

Better-quality saw blades commonly have several *expansion slots* cut down from the gullets and through the rim. As the blade heats up during a cut, the slots allow the blade to expand closer to the teeth instead of deeper in the body, which can cause warping. Some blades also feature *sound-dampening slots* that absorb excess vibration so the blade cuts more smoothly and quietly.

The business end of any blade is its teeth. You'll find that *carbide teeth* have replaced steel teeth on virtually every saw blade made these days. This is because carbide, a mixture of tungsten carbide particles and a cobalt binder, holds an edge 40 to 50 times longer than steel teeth. The downside to carbide is that it is brittle, which means carbide teeth will chip or break if the teeth strike something harder than wood (see photo, bottom left). And unlike plain steel teeth, the complex tooth grinds found on many carbide teeth are almost impossible to sharpen yourself.

Carbide teeth are brazed into pockets in the blade shoulders, ground to the proper profile with diamond abrasives, then smoothed until the teeth are razor sharp. Depending on the blade, the shoulders will hold the carbide teeth so they pitch forward or backward, forming an angle relative to an imaginary line that extends from the center of the arbor hole. This angle, known as *hook*, is considered *positive* if the teeth pitch forward and *negative* if the teeth have no angle or tip backward on the shoulders (see illustration, page 57). Blades made for rip-cutting have a positive hook angle, so the teeth remove larger chips and cut more quickly. Blades for cutting laminate and particleboard, on the other hand, have negative hook angles so the teeth scrape, rather than chip or shear, what they cut. Negative-hook blades cut more slowly and produce smaller chips, but they leave glass-smooth cuts with minimal chipping and tearout.

KNOW THE GRINDS

Carbide teeth are ground across the top in one of three ways: *flat, beveled* to the left or right, or *chamfered* on both tips. The type of grind affects how the teeth cut. By combining these three tooth grinds in different patterns, blades can be classified according to the following four grind configurations.

Flat Top Grind (FT)

FT grinds are used primarily on blades intended for rip-cutting. Each blade tooth is ground flat across the top, so it acts like a miniature chisel, chipping away material in the saw kerf. Blades with FT grinds typically have a positive hook angle of around 20° and deep gullets between the teeth.

Alternate Top Bevel (ATB)

ATB blade teeth bevel to the left and right in an alternating pattern. Rather than chip as they cut, the teeth shear wood like tiny knives. ATB blades are best suited for making smooth crosscuts, but they make adequate rip cuts as well. Bevel angles on ATB teeth will vary, ranging from 5° to 40°. Steeper, narrower bevels will cut more cleanly than shallower, shorter bevels, but the teeth will dull more quickly.

Alternate Top Bevel and Raker (ATB–R)

The teeth on ATB–R blades are configured into groups of two to four beveled teeth, followed by one raker tooth. Blades with this hybrid tooth design are commonly called *combination blades*, and they are designed to perform both rip-cutting and crosscutting functions. Combination blades are a good choice if you only buy one blade for your saw. They won't perform either rip cuts or crosscuts quite as well as blades designed only for these purposes, but they will do a decent job with either cut. If you use only a combination blade, plan on smoothing your cut edges with a hand plane or on the jointer.

Triple-Chip Grind (TC)

Half of the teeth on TC ground blades are chamfered on both corners, and the other half consists of flat-top teeth. The chamfered teeth clear out most of the waste in the saw kerf, and the flat-top teeth clean up the corners. Most TC blades have a negative hook angle in order to make smooth cuts in composite materials like MDF, melamine, and laminate. They can also cut plastic and non-ferrous metal.

Blade Storage Cabinet

Lip

Base

Store your blades in a sturdy, accessible container, like the blade cabinet shown above. This simple design employs ¼" plywood dividers to form separate compartments for each blade. If you build a similar cabinet for your blades, keep the blades from rolling out by tipping the base backward slightly, and fasten a lip along the front edge.

EVALUATING CARBIDE-TOOTH BLADE QUALITY

Here are a few pointers to keep in mind next time you are blade shopping for your table saw. Don't assume that you must buy the most expensive blades. If you are a general-purpose woodworker, buy blades that fall in the middle price ranges and have the following characteristics:

1. Blade bodies should have fine grind marks that radiate out in concentric rings toward the rim. Grinding shows that the blade was flattened, balanced, and finished to a specific thickness. Hold a straightedge against the blade body and shine a light behind it. Only the slightest sliver of light (or none at all) should show between the straightedge and the blade.

2. The arbor hole should be free of burrs. Arbor holes on inferior blades are often punched but not reamed and ground smooth. Grinding ensures that the hole is centered and will fit the saw arbor exactly.

3. Inspect the brazing behind each tooth. Brazing should be ground smooth and reveal no pitting, burrs, or gaps next to the teeth.

4. Examine the carbide teeth with a magnifying glass. They should be smooth, sharp, and polished on all edges. Buy blades with the thickest carbide teeth you can find; thicker teeth can be sharpened more times before they must be replaced.

COMMON BLADE TYPES

Should you decide to own only one blade, make it either a combination or crosscut-style. Eventually, though, you'll want to broaden your inventory to include specialty blades. Each blade type shown here is designed to make specific kinds of cuts. Some can do double duty, but you'll get top results when you saw with the blade best suited for the task.

1. **Combination Blade.** A true "jack-of-all-trades" blade, combination blades make rip cuts and crosscuts equally well in all woods. Teeth are ATB–R grind, grouped in sets of two to four beveled teeth plus one raker. Deep gullets in front of the raker teeth remove chips quickly. Tooth counts range from 24 to 50 teeth for 10"-diameter blades.

2. **Ripping Blade.** Ripping blades cut quickly and smoothly with the wood grain, which is why they have only 18 to 24 flat-top teeth and deep gullets. A ripping blade makes poor crosscuts. These blades work best when used with a saw that has at least 2 h.p.; their aggressive hook angle and wider teeth require plenty of sawing muscle.

3. **Crosscut Blade.** Crosscut blades have higher tooth counts (up to 80 teeth on 10" blades) than either ripping or combination blades. Their shallow gullets and alternating beveled teeth produce smooth crosscuts with little or no tearout in hardwoods, softwoods, and sheet goods. They also make suitable, but slower, rip cuts.

4. **Laminate-Cutting Blade.** A good choice for cutting MDF, melamine, laminate, and plastic, laminate blades combine triple-chip ground teeth plus flat-topped rakers, all set at a negative hook angle of up to -6°. These 40- to 80-tooth blades take small bites, so the cuts will be free from both tearout and chipping.

5. **Plywood Blade.** An economical alternative to carbide-tipped crosscut and laminate blades, these blades use up to 200 tiny high-speed steel teeth to make smooth cuts in plywood. Plywood blades require slow feed rates, and they'll dull much quicker than carbides, especially when used to cut particleboard. When the blade dulls, dispose of it.

DADO BLADES

Dado blades are modified saw blades designed to cut grooves up to ¹³⁄₁₆" wide. A dado blade cuts only partway through a board (called *non-through cuts*) as opposed to cutting a board in two (called *through cuts*). Dado blades are used mainly to cut interlocking parts for wood joints. For more on using dado blades for wood joinery, see pages 104–135. Incidentally, when a dado blade cuts along the grain of a workpiece, the groove is called a *plow*, while a *dado* is the proper term for cuts made across the grain.

Dado Blade Styles

Single-Blade Wobble Dadoes

Single-blade wobble dadoes use an adjustable hub to hold the blade at a skewed angle relative to the workpiece. As it spins, the blade wobbles back and forth, shearing the groove to width. Changing the hub settings increases or decreases the amount of blade skew.

Double-Blade Wobble Dadoes

Double-blade wobble dadoes form a "V" configuration that skews both blades to the workpiece. Like single-blade wobble dadoes, an adjustable center hub changes the width of the "V" for cutting grooves of different widths.

Stacked Dado Blades

The choice of most experienced woodworkers, stacked dado blades employ two carbide-tipped outer blades that sandwich chipper blades of varying thicknesses. The width of the cut equals the thickness of the outer blades plus the combined thickness of the chippers you stack together.

Types of Dado Blades

Dado blades come in two forms: *stacked dado blades* and *wobble dadoes*. Stacked dado blades consist of a pair of carbide-tipped saw blades that sandwich one to five *chipper blades* in between. The teeth on these outside blades bevel outward, so they shear the walls of the dado and establish the width of the kerf. The chipper blades clean out the waste between the walls. Most chippers look more like lawnmower blades than saw blades, with one carbide tooth on each end of a rectangular blade body. Some dado-blade sets have cross-shaped chippers instead, with four teeth.

Stacked dado-blades are so named because the outer blades and chippers are stacked on the saw arbor in a configuration that matches the thickness of the cut you need. The chippers vary in thickness, from $\frac{1}{16}$" to $\frac{1}{4}$". By switching the combination of chippers, stacked dado blades can cut dadoes or plows ranging from $\frac{1}{4}$" to $\frac{13}{16}$" in increments of $\frac{1}{16}$". For instance, to cut a $\frac{7}{16}$" groove, the stack could consist of the two outer saw blades, which typically have $\frac{1}{8}$" kerfs, plus one $\frac{1}{8}$" and one $\frac{1}{16}$" chipper blade.

When installing a stacked dado blade, it's important to stack the blades together so the chipper teeth adjacent to the outer blades are positioned within the gullet areas of the outer blades (see inset photo). This way, the teeth will overlap across the kerf width and remove all of the waste in the cut. Arrange the chippers so that they radiate around the circumference of the outer blades to balance the blades on the arbor.

Single-blade wobble dadoes cut grooves of various widths by adjusting a center hub. The hub is marked for different cutting widths, but these are rough estimates at best. Make cuts on a scrap piece first to verify the blade setting.

Double-blade wobble dadoes also are adjustable by twisting a center hub. Turning the hub changes the width of the "V" shape formed by the blades. The blade shown here also has references for determining blade height and actual cutting width.

Wobble dadoes operate on a completely different principle than stacked dado blades. Instead of two outer blades and a series of chippers in between, most wobble dadoes are simply a bevel-toothed blade mounted on a wedge-shaped hub. As the dado spins, it wobbles back and forth on its axis, shearing away a swath of material across the width of the cut. The center hub is adjustable and marked for different kerf width settings. By turning the hub, you change the pitch of the blade on the saw arbor.

A second, less common wobble-dado style consists of two bevel-toothed blades that form a "V" configuration on the arbor. Like their single-blade cousins, two-blade wobble dadoes can be adjusted to cut dadoes of various widths by adjusting a wedge-shaped center hub. As you turn the hub, the distance between the blades at the top of the "V" narrows or widens.

Choosing a Dado Blade

Stacked dado blades are more expensive than either wobble-style, and the cost among stacked dado blades will vary. More teeth or chippers generally raises the price. Despite the cost, stacked dadoes tend to produce smoother cuts than either single- or double-blade wobble dadoes. A top-quality stacked dado will cut dadoes with flat, square walls, a flat bottom, and minimal tearout on the surface of the workpiece. How well a stacked dado blade cuts will depend on how closely the diameter of the chippers matches the diameter of the outside blades, as well as the sharpness of the blade teeth.

Wobble dadoes are the least expensive type to buy, selling at about half the price of a stacked dado. Good-quality models can cut quite cleanly with minimal tearout. However, wobble dadoes produce cuts that slope at the bottom at

most settings because the blade is cutting at an angle to the workpiece rather than straight on. The amount of slope will change as you adjust the wobble dado to different width settings. Double-blade wobble dadoes are capable of cutting dadoes that are closer to flat across the bottom, but most settings create bottoms that are "dished" in or out; this is a by-product of sawing with skewed blades.

An uneven-bottomed dado is fine if the bottom of the cut doesn't have to show, or if you clean up and square the cut with a router or chisel afterward. If you plan to use a dado blade for "hogging" out material rather than for finer finished joint cutting, a wobble-style dado will suit you well.

Which Size is Right for Your Saw?

Both stacked and wobble-style dado blades are sold in 6"-, 8"-, and 10"-diameters. Manufacturers recommend using a dado blade with a diameter 2" smaller than the maximum blade size for your saw. The smaller size recommendation is due to the fact that dado blades are heavy and create more stress on the saw arbor as they spin. Wobble-style dadoes produce considerable vibration as well, which must be soaked up by the saw, or they'll transfer into rough cuts. Be wary of using a dado blade on jobsite saws with motors rated at less than 1½ h.p. Refer to your owner's manual to see if a dado blade is a suitable accessory for your saw.

MAKING BASIC CUTS

Table saws are not difficult to use successfully and safely as long as you follow a few basic guidelines. First, you need to choose your lumber carefully. Wood continues to change shape long after it's cut from the tree, due to fluctuations in moisture content and other internal stresses in the wood. Evaluate the condition of each piece of wood you plan to cut before you cut it. You may need to correct for various kinds of wood defects with other tools like jointers or portable saws before it's suitable for the table saw.

Second, know exactly which kind of cut you are making and have a specific plan in mind. Table saws essentially cut wood in two ways: along the grain (called *ripping*) and across the grain (called *crosscutting*). The procedures for making these two kinds of cuts are quite different. Be clear about these differences and exercise safe technique.

Third, take into account the bulkiness and proportions of your workpieces so you can negotiate them safely across the saw table. Your goal is to maintain control over yourself, the workpiece, and the saw at all times to avoid accidents and mistakes.

Turning on the saw and passing wood through the blade is actually the quickest step of the cutting process. Choosing and measuring up your stock, setting a fence, clamping or supporting your workpieces and reviewing a cut in your mind is most time-consuming, but these steps are all necessary parts of a process that will ensure accurate, safe sawing.

Caution

Some of the photographs you'll see throughout this book show the blade guard and splitter or riving knife removed. This was done to make the information in the photographs easier to see. To prevent injury, use a working blade guard and a splitter or riving knife with anti-kickback pawls whenever the type of cut being made allows it.

BEFORE YOU START CUTTING

Cutting wood involves more than just tuning up a table saw and installing a sharp blade. You also need to start with sound lumber that can be made flat and square on at least one face and one edge or end. Chances are, most lumber that ends up in your shop is neither straight nor flat. It also may have a number of other natural defects that are undesirable or even unsafe if they should come in contact with the blade. The next few pages will outline a few important properties of wood you should know, identify lumber defects to look out for, and then help you evaluate how to correct these problems to create the best possible workpieces for sawing.

Cutting boards is an activity that involves two main players: the lumber and the saw. Selecting good lumber or salvaging what you can from inferior boards is as important to your cutting successes as using a sharp blade on a well-tuned saw. Workpieces should be dry, free of serious defects, and flat and square to ensure accurate, safe results when you make your final cuts.

Understanding Wood Movement

Since wood once served a structural purpose for the tree, it retains a certain amount of internal stress that may resist your efforts to machine it straight and flat. The wood may have held up a heavy branch or come from a part of the trunk that twisted or curved in some way. A single board may be cut from two different regions of the tree trunk, and both regions will influence how the board reacts when you cut it.

Wood also acts like a sponge to water. Until you apply a protective topcoat to slow down the rate of change, expect your lumber to expand when the humidity rises in damp summer months and contract when the air dries out in fall and winter. Expansion happens most noticeably across the width of a board, perpendicular to the grain direction. If a board doesn't absorb and release moisture evenly, it will begin to distort or even split in extreme cases. To keep your lumber as stable as possible, store it in a place where levels of humidity remain relatively constant. Keep as much of the surface area of each board exposed as you can by stacking boards with narrow wood spacers in between, to improve the airflow around each board.

Testing Moisture Content

Lumber varies widely in terms of moisture content. The amount of moisture a board contains will affect how well the board cuts as well as how true and flat it will stay from then on. Wet lumber is more difficult to cut; once cut, it can split or warp as it dries.

The ideal moisture content for furniture lumber is around 8% to 10%. The only way you can accurately test for moisture content is to use a calibrated moisture meter, available through woodworking catalogs. Most meters have a pair of sharp sensor probes that you press into the board. A readout scale on the meter tells you the moisture percentage. Moisture meters can be expensive, but you may want to invest in one if the humidity in your shop fluctuates widely. You'll also want to test lumber for moisture content if you buy it roughsawn directly from a sawmill. Kiln-dried lumber is best, but air-dried lumber is also fine for woodworking if it's been dried for at least a year. Ask your supplier.

It's best to set aside lumber with a moisture content over 12% and allow it to air dry before you cut it.

A moisture meter will tell you immediately the moisture content of a board. The red glowing light on this meter indicates the moisture content in this board to be 10%, an acceptable level for cutting and for project use.

Identifying and Evaluating Wood Defects

In any stack of lumber, you'll find four kinds of natural distortion occurring: twisting, bowing, crooking, and cupping (see below). If a board's distortions are slight enough that you can't detect them at first glance, you can probably flatten and square it fairly easily using the techniques outlined on the next few pages. However, it may be impossible to salvage much worthwhile wood from severely distorted lumber. Keep in mind that even if you correct one or another kind of distortion, it could reemerge over time. Evaluate how important a piece of distorted wood is to your project. You may want to discard it for a more suitable piece instead.

Other wood defects to identify are *checks* (cracks that usually begin on the ends of boards), *loose knots, pockets of sap, spalting, or tunnels* left by wood-boring insects. You may even find boards with bark still attached (called *wane*), if you buy your lumber directly from the sawmill. It's impossible to avoid all lumber with defects, and some, like tight-grained knots, can even be desirable. Generally, however, you should cut off these problem areas first to start with the soundest workpieces you can.

Types of Wood Distortion

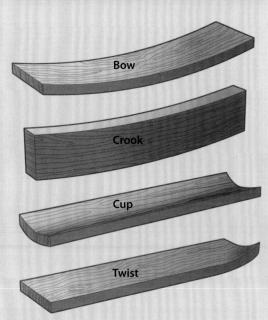

Bow

Crook

Cup

Twist

Boards distort in four ways, depending upon internal stresses as well as how the board absorbs and releases moisture. Wood that *bows* is flat across its width but the faces curve lengthwise. A *crooked* board is flat across the face but curves along the edges in one direction or the other, like the rocker on a rocking chair. *Cupping* is occurring when a board is flat along its edges but curls across its width. *Twist* is the condition where one or more corners of a board turn up or down when you lay the board on a flat surface. It's fairly common to find a couple of these distortions happening in the same board. The more pronounced the distortions are in a board, the harder it will be to salvage much useful lumber.

Spalting is a gray to green permanent discoloration of the wood caused by fungal growth.

Knots should be avoided when purchasing hardwoods, but in some cases you may be able to work around them.

Checking is a splitting of the wood grain at the end of the board. It can be cut off, but it may indicate that the board was not dried properly. If so, the board may actually begin to split again.

Salvaging Lumber

In many instances, you can salvage distorted or defective lumber by cutting, jointing, and planing problem areas away, creating smaller pieces of lumber with flat faces and edges. Generally, the easiest and safest way to cut down irregular boards is to use a portable circular saw, jigsaw, or handsaw (see photo, top). The small base of a portable saw can follow the irregularities of a twisted board far better than a large, flat table saw table can. Use the portable saw as a resizing tool, then cut your finished edges on the table saw.

If you choose to cut defective lumber down first on the table saw, start with at least one flat edge for guiding against the rip fence or with the miter gauge. Flatten the edge with a hand plane or on a power jointer. If you don't own a jointer, you can flatten the edges of a crooked or waney board on your table saw by overlapping one undesirable edge with another board that has a flat edge. Fasten the boards together with finish nails or small screws. Run the flat edge of the tacked-on board against the saw fence to trim off the opposite, rough edge on the defective workpiece (see photo, middle). Separate the boards and run the cut edge of the workpiece against the rip fence to square the board.

When cutting bowed lumber on the table saw, set the flatter face down to minimize rocking as you cut. If a board is cupped, it's best to make rip cuts with the concave side facing up (see photo, bottom). Ripping with the convex side facing up will cause the board to flatten out against the blade as the cut progresses, which could bind the wood between the fence and blade and cause it to kick back. If the board is sufficiently thick, you also may be able to correct for cupping by knocking off high areas first with a hand plane or on the jointer to flatten it, then running the board through a surface planer to smooth the faces. Simply running a cupped board through a planer won't eliminate cupping. A planer will temporarily flatten the board as it passes through the feed rollers, but it cannot remove the curling stresses present in the wood.

As far as other imperfections are concerned, cut off areas with loose knots or splits. Cut well around soft pockets of sap to keep from gumming up your blade. If you are using wood recycled from other projects, always inspect it carefully for nails, staples, or other construction debris before you saw. Hitting a metal fastener will invariably damage your saw blade.

Trim off split ends, loose knots, and other wood defects with a portable saw before cutting a problem board on the table saw.

Flatten the edges of crooked lumber or remove bark-covered (waney) edges by attaching a flat scrap over one rough edge. Run the flat edge against the rip fence. Then, remove the flat scrap and flip the board around to cut the other edge flat.

Cut cupped lumber with the concave side facing up. In severe cupping situations, as with this board, saw the board into smaller flat pieces. Keep the board from rocking as you cut it, to reduce the chances of kickback.

Procedure for Squaring Up Stock

Flatten and smooth board faces. Once you've cut away defective areas and flattened high spots with a hand plane or jointer, smooth both board faces by running it through a surface planer, alternating the board faces with each pass.

Flatten one edge of your stock with a planed face against the jointer fence. Then guide this flat edge against the rip fence on the table saw to cut the board to width. Joint this cut edge smooth, if necessary, to remove saw marks.

Setting Up for the Cut

Before cutting a board, you'll need to make some decisions about how to set up the table saw and related safety equipment to accomplish your task most effectively and safely. Ultimately, your goal should be to keep your motions fluid and controlled as you push wood through the saw blade. Fluid, steady motions will produce smooth, accurate cuts and minimize your chances for injury. It will also make cutting wood an enjoyable part of woodworking. To this end, here are a few suggestions to follow as you prepare for any cut:

- **Start with flat, square stock that is free from defects and is reasonably dry.** (See the above photos for more information on squaring up stock.)
- **Give yourself plenty of room around the saw.** Route the power cord so it won't pose a tripping hazard. Allow yourself 4' to 6' of room to the left and to the right of the extension wings and 8' to 10' in front of and behind the saw. The amount of clear space you'll need will vary, depending upon the dimensions of the workpiece you're cutting and whether or not you'll be using extra outfeed support devices as you cut. The goal here is to create a workspace that allows you to keep your attention focused entirely on what's going on at the saw table, not on the side stepping you might need to do to get around the pile of scraps you didn't clear off the floor from the last few cuts. Nothing should break your concentration or obstruct the path of your workpiece while you are making a cut.

Keep the Sawing Area Clear

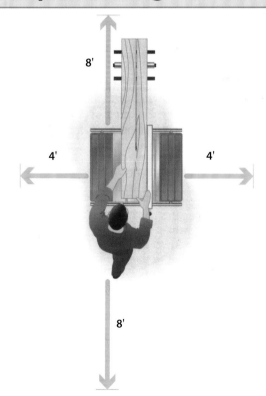

Make room for sawing. Allow 4' of clear space on each side of the saw and 8' to 10' in front and in back so you'll have plenty of room to work.

- **Choose the right blade for the material you are cutting.** Refer to the chapter on "Choosing Saw Blades," pages 56–62, to match your cutting tasks with the proper blade.

- **Be sure the saw guard is in place and the splitter or riving knife is aligned squarely behind the blade.** For more on adjusting them, see page 50.
- **Clamp featherboards to the saw table when necessary.** Featherboards should be positioned just in front of the blade to guide the workpiece into the blade without binding against it. Never use featherboards or hold-downs on the outfeed (back) side of the blade when you are cutting a board into two pieces. Doing this could close up the saw kerf behind the blade and invite kickback. Set up your featherboards so the workpiece slides smoothly beside or beneath the featherboard without dragging or binding, but the featherboard keeps the board from backing out of the cut when you pull on the board (see next page for an example of proper featherboard setup).
- **Set the blade height to produce the smoothest cut possible without compromising your safety.** The ideal blade height is a matter of debate among professional woodworkers.

Setting Blade Height for Making Through Cuts

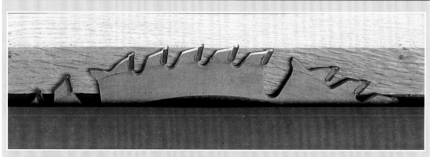

When the blade is set low, so that only the tips of the teeth protrude above the board, all of the blade teeth above the saw table remain in the workpiece during the cut. With more teeth in the board, the blade develops more heat, which results in burned edges and, some will argue, an increased risk for kickback.

With the blade raised to full height, only a few teeth contact the workpiece at the front and back of the blade, resulting in a much cooler, and generally cleaner, cut. However, you are at greatest risk of blade exposure, especially if you saw without a guard.

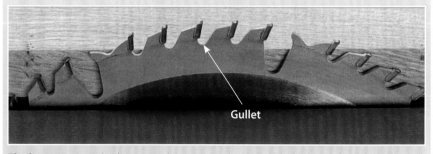

Gullet

The best compromise between clean cuts and minimal blade exposure is to set the blade just high enough so the gullets between the teeth clear the top of the board (about ¼ to ⅜" on most blades).

The safest blade is one with a minimal number of teeth protruding above the workpiece (about ⅛" to ¼" exposure). However, low blade height means more teeth remain in the cut for longer periods of time, which leads to more friction in the saw kerf, resulting in burn marks on cut edges. As you crank the blade higher, notice that fewer teeth on the front and back edge of the blade are actually in the cut, which means the blade will heat up less and produce a smoother cut, but your risk of getting cut increases. Experiment with different blade height settings to see which works best for your work and gives you the most peace of mind.

If higher blade settings make you nervous, saw at a lower blade height, cutting just outside of your layout line. Then run the sawn edge over a jointer to remove

Use featherboards on top of or beside a workpiece to hold it securely against the saw table and the fence. Featherboards are designed to flex with the direction of feed but bind against the workpiece if the blade should attempt to force it backward and out of the cut. Install featherboards on the infeed side of the blade when you are cutting a workpiece in two.

any saw marks that a lower blade might leave on the workpiece. This way, you'll have the best of both worlds in the end—a safer blade setting for sawing and enough of a waste edge to joint smooth.

- **Keep a push stick within arm's reach at all times.** Use a push stick to guide the workpiece whenever your hands come within 6" of the blade. If you'll be ripping narrow stock between the blade and the rip fence, be sure to choose a push stick narrow enough to fit between the blade and the workpiece.

- **Set up outfeed support for long workpieces.** When making long rip cuts, you'll need a roller stand or worktable set up behind the saw to support the workpiece and waste piece as they exit the back of the saw. For long crosscuts, set up workpiece support alongside the saw, just beyond the left or right extension wing. Be sure the height for outfeed support stands is even with or slightly below the surface of the saw table so workpieces won't hang up on them during a cut. There are many different aftermarket outfeed support options available, including outfeed and extension tables that attach directly to the saw. You can also make your own (see pages 160–212).

Whichever outfeed device you choose, it should have a wide, firm stance on the floor so it will be stable in all directions and not prone to tip or shift away from the saw as you cut. You could also use a helper to support workpieces as they leave the saw (see page 98).

Making Blade Changes

Installing a saw blade is a simple operation, and you'll do it many times, especially if you use the same saw for all your cutting operations. First, unplug the saw. Remove the throat plate and raise the blade to its full height. Wedge a piece of scrap wood into one of the blade gullets to keep the blade from turning as you loosen the arbor nut. Examine the threads on the arbor to see whether they are cut clockwise or counterclockwise on the arbor, then use either the arbor nut wrench that came with the saw or an adjustable wrench to loosen the arbor nut. (Most table saws have left-handed threads on the arbor, so you'll turn the arbor nut clockwise to loosen it.) Remove the nut, being careful not to drop it into the saw base or onto the floor. Pull the arbor washer off and, with two hands, slide the blade off the arbor. Store the blade in a dry place with its teeth protected. Install a blade by reversing the removal procedure, but be sure to clean any sawdust deposits off the arbor flange and threads first. Tighten the nut firmly, but not so tight that it makes removal difficult later. Reinstall the throat plate and adjust it so it sits flush with the surface of the saw table.

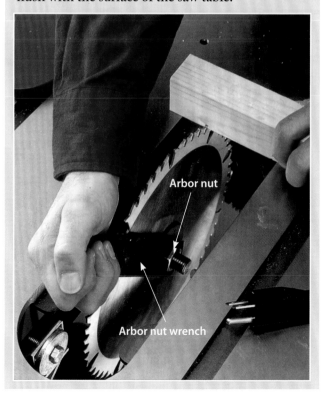

Arbor nut

Arbor nut wrench

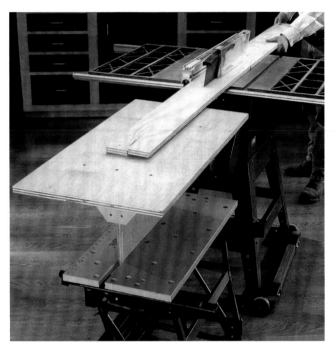

Provide a stable outfeed surface to support your workpiece when it extends over the side or back of the saw. Plans for building this handy outfeed table are found on pages 160–161.

Roller stands are another option for outfeed support. Workpieces glide over ball bearings on the one shown here. Anchor the base with a weight to keep it from shifting.

Table Saw History

Self-Feeding Ripping Saw

This saw, manufactured in 1902, served as a production ripping saw for planing mills; furniture, wheel, wagon, buggy, and piano factories; car, railway and locomotive shops; navy yards; and arsenals. As shown in this engraving, the saw could be outfitted with a belt-driven, two-speed self-feeding attachment, which used a toothed disk in front of the blade and either a smooth or fluted roller on the outfeed side.

Built mostly of cast iron, the saw featured a patented saw table that raised and lowered on a pair of screw gears to change blade height, rather than tilting the saw arbor. The 1⅜"-diameter steel blade arbor could accept one or multiple saw blades up to 22" in diameter. Driven by a centralized belt system in the factory, the arbor spun at 2,400 revolutions per minute, about half the speed of table saws built today. Notice the long rip fence and the fence scale mounted above the fence on the right side of the saw.

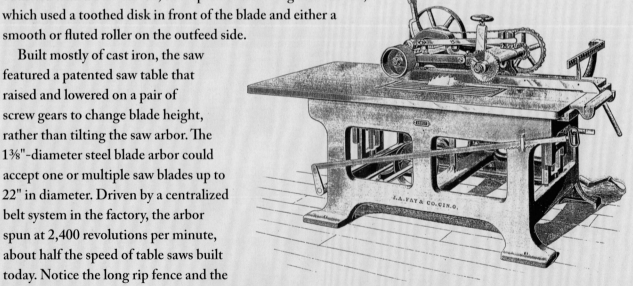

J. A. Fay & Co. No. 2½ Self-Feeding Ripping Saw (circa 1902)

MAKING RIP CUTS

To *rip-cut* (or *rip*) a board means to cut it lengthwise, following the direction of the grain. Of the basic saw cuts you can make on the table saw, you'll probably spend more time ripping workpieces than performing any other kind of cut. This is because the table saw, with its self-aligning rip fence, is an exceptional tool for making rapid, smooth rip cuts.

Rip cuts can be made with portable saws, but the process is more time-consuming. It's also tricky to rip-cut accurately unless you have a steady hand to guide the saw or clamp a straightedge to your work. You'll need to set up workhorses or clamp your workpiece to a workbench—a task you won't need to do with a table saw.

Table saws, on the other hand, use the rip fence as a guide for rip-cutting, so the accuracy of the cut doesn't depend on how steadily you can hold and maneuver the saw along your layout line. Success of the cut, however, will still depend on how smoothly you guide the workpiece through the blade.

Setting the Rip Fence

To make accurate and safe rip cuts, the rip fence must be parallel to the blade (see "Understanding Blade Heel," page 44). Begin setting up for the cut by sliding the fence along the fence rail until its distance from the blade roughly matches the width of the board you need to cut. Under most circumstances, the finished workpiece is cut from the portion of the board between the rip fence and the blade, and the waste piece is the section that falls away from the blade on the other side when you finish your cut.

With the saw unplugged, use a tape measure or rule to set the exact distance you need between the fence and the blade. Do this by measuring from the inside edge of a blade tooth (the tooth edge that faces the rip fence) to the face of the fence. Tap the fence into place. Don't make the mistake of measuring from the "outside" edge of a tooth, or your finished workpiece will be too narrow. Lock the fence into position and check the width setting again, to be sure the distance hasn't changed.

Raise the blade to a height sufficient to cut through the workpiece in one pass. How much higher than the workpiece you set the blade is a judgment call that depends on your own comfort level. (See page 68 to learn more about the pros and cons of high versus low blade settings.) If you need to rip stock thicker than the maximum blade height of your saw, see "Ripping Thick Boards," page 79.

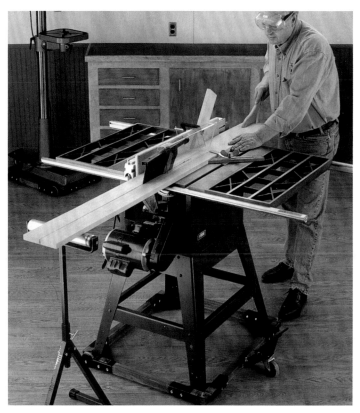

Rip cuts. When you cut a board along the grain, you are making a rip cut. You'll likely make more rip cuts than any other kind of cut on a table saw.

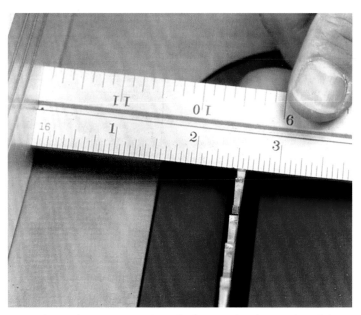

To set the rip fence, measure from the fence to the closest edge of the blade. This distance should match the exact width of the workpiece you need. For instance, the setup in this photo will cut a workpiece that is 2½" wide.

Squaring the Rip Fence

If your rip fence tends to shift out of parallel when you lock it down, which happens with many rip fences, try this simple tip: Before you lock the fence, press your palm against the center of the fence body that rides along the front fence rail to square the fence on the fence rail, then lock down the fence.

Safe Hand Zones for Rip Cuts

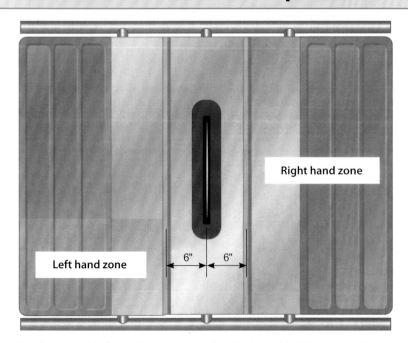

Right hand zone

Left hand zone

6"   6"

Hand zones. With the rip fence positioned to the right of the blade, your right-hand zone extends across the depth of the saw table for pushing workpieces completely past the blade. Your left hand holds workpieces in place against the table as soon as they are cut free, so there's no need to extend it past the infeed edge of the blade. Neither hand should come within 6" of the blade. Always use push sticks instead when you are this close to the blade.

Start the Cut

Typically, rip-cutting is done with the fence on the right-hand side of the blade. In these situations, stand in front of the saw to the left of the blade with your left foot touching the left corner or leg of the saw base. The closer your body is to the saw, the less you'll need to reach over the blade as you complete the rip cut, which will help you maintain your center of balance throughout the process. If your workpiece is short enough, rest your waist against the front fence rail for added support. Stand far enough to the left of the blade so your waist clears the width of the workpiece. This way, if the saw should kick the workpiece back out of the blade, it is less likely to strike your body. However, stand as close as you can to the side of the workpiece so your right arm is aligned with the fence and the blade as you make the cut.

Tip

You can judge your feed rate by the sound of the motor as you cut; try to keep the motor cutting at an even pitch. You'll gain a "feeling" for the appropriate feed rate the more you use your saw. Slow down your feed rate if the motor begins to bog down, especially when cutting particularly hard, wet, or thick workpieces.

Tip

It's natural to want to glance at the blade while you cut, but avoid the temptation—it won't tell you much about how the cut is progressing anyway.

How to Make a Standard Rip Cut

Start rip cuts with your left foot against the front left corner of the saw base. As you cut, apply three forces simultaneously to the workpiece—forward pressure with your right hand to feed the work, and downward and side pressure with your left hand to keep the workpiece flat on the table and snug against the rip fence. Watch the workpiece to make sure it is making contact with the rip fence. At no time should the workpiece stray away from the fence. Have a push stick handy whenever a workpiece requires your hand to be closer than 6" to the blade.

Continue steadily feeding the workpiece through the blade with your right hand as the end of the board approaches the front edge of the saw table. For workpieces wider than 6", use your right hand to continue guiding the board through the blade. If the workpiece is less than 6" wide, use a push stick instead of your right hand. Guide the board with your left hand and reach for a push stick with your right (see illustration, previous page). Position the push stick securely over the end of the board, and use it to continue pushing the workpiece past the blade.

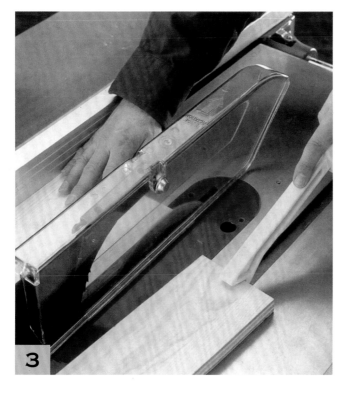

When the blade finally separates the board in two, hold the waste clear of the blade with your left hand or a second push stick while you feed the other part of the workpiece clear past the blade with your right hand/push stick. Then clear the waste piece off the saw table by sliding it away from the blade with your left hand or a push stick.

Troubleshooting Chart for Rip Cuts

Problem: Workpiece stops on the saw while making a cut

- **Possible causes:** Typically a workpiece will hang up in a cut because the splitter or riving knife are not in line with the blade, so the splitter or riving knife catches the leading edge of the board. If your workpiece is thick, the board may be hitting the mounting posts that hold the guard to the splitter or riving knife; otherwise, the workpiece may be catching on the back edge of the throat plate.

- **Solutions:** Inspect and realign the guard and splitter or riving knife (see page 50). Adjust the throat plate (see page 41).

Problem: Workpiece veers away from the fence during a cut

- **Possible causes:** Workpiece edges could be crooked (see page 65). Or, you could be steering the board with your right hand rather than just pushing it as you make the cut.

- **Solutions:** Always work with flat, square stock. Install a featherboard and try another cut. If the problem persists, the blade is likely heeling on the outfeed side. Align the blade with the miter slots (see pages 44–45).

Problem: Workpiece begins to smoke during a cut

- **Possible causes:** Wood can burn for a variety of reasons as you saw. Investigate these possible culprits: Is your blade dirty or dull? Is the saw kerf closing up around the blade on the outfeed side? Is your rip fence parallel to the blade?

- **Solutions:** Clean your blade (see page 55). Install a splitter or riving knife if you aren't already using one to keep the saw kerf open behind the blade. Adjust the rip fence so it is parallel to the miter slots and blade (see page 48).

Problem: Blade slows or stops during a cut

- **Possible causes:** Stalling usually indicates that the saw kerf is closing up or you are sawing with a dirty or dull blade. If you are using a small saw to cut thick hardwood or wet or particularly resinous wood like fir, the wood may simply be taxing the limits or your saw. If you are using an extension cord, feel it. A warm cord or plug indicates that you may be using an undersized cord, and the motor is starving for power under load.

- **Solutions:** Be sure to use a splitter or riving knife. Slow down your feed rate as you push wood through the blade to give the saw more time to cut and the blade to evacuate chips from the kerf. Clean your blade and/or have it professionally sharpened. Or you may want to switch to a thin-kerf blade if you have a small saw (see page 58). Use the proper gauge extension cord to power your saw (see page 30).

Problem: Workpiece lifts off the saw table near the back of the blade during a cut

- **Possible causes:** The workpiece could be bowing (see page 65). Or the kerf could be closing up behind the blade, and the blade is actually lifting the workpiece off the saw table. If you are using a featherboard, check to see that the clamps haven't shifted out of place or loosened their grip on the workpiece.

- **Solutions:** Inspect your workpiece by sighting down its length to see if it is bowing. Check your rip fence's alignment to the blade to be sure the blade isn't heeling (see page 44). Tighten the clamps that hold the featherboard in position, if you are using one. In any case, it is imperative that workpieces stay snug against the saw table during a cut to avoid kickback.

Ripping Long Boards

When rip-cutting boards that are longer than about 4', you won't be able to begin the cut with your body next to the saw table. In these cases, first set up an outfeed support device behind the saw.

Rest the front end of the board on the front edge of the saw table. Start the saw, move to the back of the board on the left-hand side and tilt it up, using the right- and left-hand positions for normal ripping. To start the rip, you'll want to keep the front edge of the board flat on the table as the blade engages it. However, this can be difficult to do when you're standing a few feet back from the saw. To compensate, lift the back edge of the board higher than the saw table to plant the front edge of the board firmly against the table. Then begin the cut, lowering the board so it rests on the saw table and walking with the board as you feed it through the blade with your right hand. Keep your eyes on the fence, and apply diagonal pressure with your left hand so the board stays snug with the fence. It will be trickier to keep long stock against the fence, so slow down your feed rate if necessary and make minor body position adjustments the minute you see any gap developing between the fence and the workpiece. Once you reach the saw table, position your left foot against the left corner of the saw base and use push sticks as necessary to complete the cut and clear away the waste piece.

Ripping Short Boards

Ripping short boards (less than 10" or so in length) can be hazardous, because you'll need to feed and support the workpiece close to the blade from start to finish. It can be tricky or impossible to work push sticks close enough into the blade guard to guide the workpiece, so you may be inclined to remove the guard altogether, which will make short rips even more dangerous. What's more, if your workpiece isn't longer than the blade, you'll complete the cut before the leading edge of the workpiece is even past the blade, which renders the splitter or riving knife useless. The best option is to try to make your rip cut from a longer piece of stock, then crosscut the shorter length you need from the longer stock. This way you'll maintain full control over the workpiece without compromising your safety.

When ripping long boards, tip the back edge of the board up so it is slightly higher than the saw table. Doing this will press the leading edge of the board firmly down on the saw table. Lower the back edge of the board as the cut progresses.

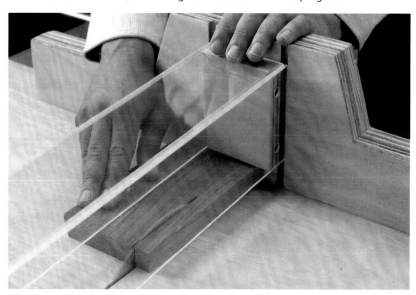

The safest technique for ripping short boards is to guide them in a crosscut sled rather than sliding them along the saw table against the rip fence. Be sure the crosscut sled you use is equipped with a blade guard. For more on crosscut sleds, see page 91.

At times, ripping short lengths of stock is unavoidable. In these circumstances, the safest course of action is to make the rip cut with a short workpiece immobilized in a crosscut sled (see bottom photo, above). The advantage to using a sled in this situation is that you can hold the workpiece steady instead of having to slide it along the saw table against the fence. Be sure to hold the workpiece securely in the sled as you make the cut. Use a push pad or push stick as a hold-down if necessary, especially if your crosscut sled has no blade guard.

Ripping Narrow Workpieces

Many woodworking projects will require you to rip strips of narrow stock into parts such as glass retainer strips, filler pieces, or trim molding. If you have only one part to make, adjust the rip fence so the finished workpiece is actually the part that falls away from the blade when you complete the cut. This way, you'll keep the wider part of the stock between the rip fence and the blade, which is safer for your hands and a steadier means of guiding the workpiece than using a narrow push stick.

When rip-cutting a single narrow workpiece from a wider board, the narrow strip should be on the opposite side of the blade from the rip fence. With this setup, you'll keep the wider portion of the board between the blade and the rip fence to allow more room for your hand or a push stick.

Options for Making Narrow Repetitive Rip Cuts

To rip a series of narrow workpieces, set the distance between the blade and the rip fence to match the intended width of the workpieces you need. Use a narrow push stick to guide the pieces along the rip fence and past the blade.

Another option for ripping narrow strips is to use a jig with a notched cutout to support the workpiece. The notched area of the jig should match the width of the workpieces you are ripping. Support the outside edge of the workpiece as you cut.

If you need to rip a number of identical strips of narrow stock, the procedure to use will depend on how wide the final workpieces need to be. If you can set the fence far enough from the blade to fit a push stick in between, then set up your rip cuts so the "inside" piece (the part between the blade and rip fence) matches the width of the pieces you need to make. The "waste" piece will constitute the remainder of the stock you're ripping from on the other side of the blade. Use a narrow push stick to guide workpieces through the blade (see bottom left photo, previous page).

The procedure changes if your strips are so narrow that the space between the rip fence and the blade guard doesn't leave enough room for the width of a narrow push stick. In these cases, narrow rip cuts are still possible, but you'll have to build a simple jig with a notched cutout that matches the width of the strips you need (see bottom right photo, previous page). The width of the jig isn't important, so long as it is wide enough to give you room between the fence and the blade to safely negotiate your hand or a push stick.

To set up the cuts, rest the jig against the rip fence so the notched side faces the blade, and adjust the rip fence until the blade touches the outside edge of the jig's notch. The blade will cut flush against the notch without cutting it off. Set your workpiece in the notch and make your rip cuts, supporting the outside corner of the workpiece closer to you with your left hand or a push stick. As the workpiece narrows, use a push stick in your left hand to keep your hand a safe distance from the blade.

Support long, narrow rip cuts as they leave the saw table just as you would for wider, heavier workpieces. Otherwise, workpieces will tend to tip up and off the back of the saw table as soon as the board is cut in two.

Understanding and Preventing Kickback

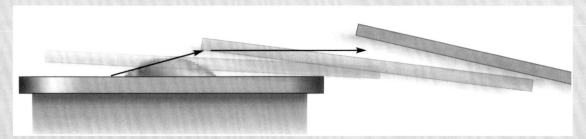

Kickback. A split second is all the time it takes for a saw blade to grab hold of a workpiece and throw it back in the direction of the saw operator. Protect yourself from kickback situations by sawing with a splitter or riving knife in place, equipped with anti-kickback pawls. Stand clear of the workpiece's path behind the saw as you are cutting.

Kickback occurs when the saw blade grabs a workpiece and propels it straight back toward the saw operator. Typically kickback happens because the workpiece binds on the outfeed (back) side of the blade, pinching the blade enough for the saw teeth to grab hold and lift the board off the saw table. A saw can eject a board in a number of cutting situations, whenever the teeth have the opportunity to grab hold of a board. However, kickbacks are most likely to happen when making rip cuts without a splitter or riving knife installed on the saw. You'll have no time to react when kickback happens, which is why splitters or riving knives are so essential to making safe rip cuts. Never make rip cuts with the splitter or riving knife removed. Stand clear of the path of your workpiece as you are making a rip cut in order to keep the board from striking your body, should kickback occur.

Using a Partial Rip Fence

It isn't always necessary or even a good idea to make rip cuts against a fence that extends from the front to the back of the saw table. When you are rip-cutting wet, green, or wavy-grained wood, the saw kerf has a tendency to open up as the workpiece leaves the blade area. If it does, a full-length rip fence can actually force the spreading workpiece back against the side of the blade, which will lead to burned edges and can result in kickback. With the exception of large sheet stock, which requires a longer bearing surface, workpieces really only need to be guided by the rip fence until they contact the blade. Some European table saws don't even come with a full-length rip fence.

To avoid the problem that can be caused by a spreading saw kerf, attach a shop-made partial fence (also called a *half fence* or *short fence*) to your rip fence. The partial fence should be made from a piece of dimensionally stable hardwood or sheet stock about 2' long. Rout a ⅝"-wide slot about 18" long and 1¼" deep along the center of the partial fence. This slot will form a recess for the heads of two countersunk carriage bolts.

Rout a ⁵⁄₁₆"-wide slot inside the first slot that penetrates all the way through the thickness of the partial fence. The purpose of this slot is to make room for the carriage bolt shafts and allow you to adjust the partial fence forward or backward along the rip fence.

Mount the partial fence to your rip fence with two ¼" carriage bolts, nuts, and washers. You may have to drill guide holes through your metal rip fence if the fence doesn't have holes already drilled for attaching auxiliary fences.

Setting the Partial Fence

Set the partial fence so the end closer to the blade extends about 1" past the blade teeth on the infeed side. Adjust the position of the partial fence whenever you raise and lower the blade, since the location of the infeed blade edge changes as you change blade height.

Build a partial fence from a length of flat, square hardwood or sheet stock. Rout a pair of grooves one inside the other lengthwise along the partial fence, and attach the partial fence to the rip fence with ¼" countersunk carriage bolts. The slots allow you to reposition the partial fence forward or backward along the rip fence as needed.

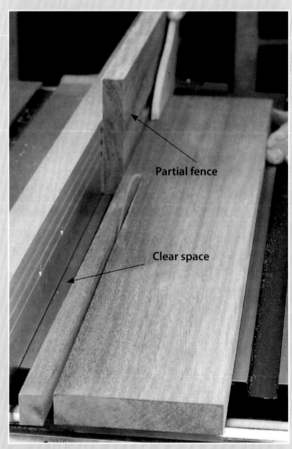

Partial fence

Clear space

The purpose of a partial fence is to provide clear space between the blade and the rip fence on the outfeed side of the blade. In cases where the saw kerf is likely to spread apart from internal stresses in the board, a partial fence will keep the workpiece from binding against the blade.

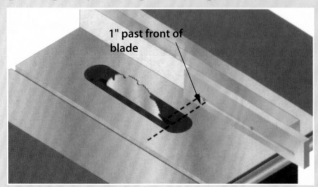

1" past front of blade

A partial fence should extend from the front edge of the rip fence to about 1" past the infeed edge of the blade.

Overarm blade guard

1

To rip stock thicker than the maximum blade height of your saw, make the first cut with the blade set to full height. Because a standard blade guard/splitter can't be used to make this cut, consider outfitting your saw with an overarm-style blade guard, as seen above. Set the rip fence so the blade will cut ½₂" or so on the waste side of the layout line. Ripping oversize stock will almost certainly leave burn marks on the wood. By sawing outside the layout lines, you can remove burn marks on the jointer or with a hand plane later.

2

Once you've completed the first rip cut, flip the workpiece end-for-end. Lower the blade so it cuts just past the top of the first saw kerf. (This way, less of the blade will be pinched if the kerf should begin to close up, a common and dangerous occurrence when you saw without a splitter.) Make the second pass. For both steps of this technique, you'll likely have to slow down your feed rate to keep the saw from bogging down in the cut.

Ripping Thick Boards

Boards that are more than about 3½" thick cannot be rip-cut in a single pass on most 10" saws. That's why most woodworkers prefer to rip thick lumber on the band saw rather than the table saw. Band saws have much thinner blades, so they create less friction and heat in the cut. Plus, band saws offer a much greater depth of cut (10" or more on most 14" band saws) than table saws.

If you don't own a band saw, you can rip thick boards on your table saw, but the procedure will require you to make two passes to complete the cut. Also, you'll have to remove the splitter on older saws if the splitter has anti-kickback pawls that would interfere with the cut. Standard guards and splitters on saws made before 2009 came as one-piece assemblies. If you have one of these older saws, you'll have to remove both the guard and the splitter, which leaves you vulnerable to both kickback and blade exposure.

The safest saws for ripping thick boards are those with a splitter or riving knife that is separate from the guard assembly. In most cases, this will mean using a saw with an aftermarket overarm guard system installed (see pages 140–142).

Table saws built after 2009 will have riving knives instead of splitters. Riving knives can remain installed on a table saw for ripping thick boards like this because the guard and anti-kickback pawls can be removed separately. It's an enhancement that improves all rip cut operations by reducing the incidence of kickback due to workpieces pinching the blade during cutting.

Tip

As you make your cuts, feed the wood through the blade as quickly as you can without bogging down the motor to minimize burning.

Ripping Thin Boards and Laminate

Most standard rip fences hug the saw table tightly enough that you can rip stock as thin as ⅛" without difficulty. However, if a gap exists between the saw table and the fence when the fence is locked in position, ripping thinner materials like plastic laminate may pose a problem. Laminate can slip into the joint between the rip fence and the saw table as you cut it.

Before you cut laminate, inspect the fit between the rip fence and saw table, with the fence locked down. If you see a gap, attach an auxiliary wood fence to the rip fence to cover the gap.

Be particularly wary if your saw fence is T-square style (see page 14), in which case the rip fence doesn't clamp to a fence rail along the back of the saw table. If using a T-square fence, attach a C-clamp around the back of the fence and the saw table to hold the fence tight to the table along its full length.

Another concern with rip-cutting laminate or other thin material is that wind generated by the spinning blade may lift the material during cutting. If the laminate starts fluttering against the blade, it will chip. Solve the problem by using a push stick with a long sole to hold the laminate tight against the saw table near the blade. For best results, use a triple-chip laminate-cutting blade or a sharp crosscut blade to cut laminate cleanly (see page 60).

Ripping Bevels

Bevel cuts are rip cuts made with the blade tilted to an angle other than 90°. Bevels are cut for a variety of purposes: to "break" a sharp edge along the length of a workpiece; to join two edges or ends together at an angle; or to cut tapered edges, such as when building the center panel on a raised-panel door.

Aside from tilting the blade, the process for making a beveled rip cut is the same as making an ordinary rip cut. However, when you tilt the blade, be sure to set the rip fence on the side of the saw table opposite the direction the blade is tilting. (For instance, on right-tilting saws, set the fence on the left side of the blade.) If the blade tilts toward the fence rather than away from it, you'll trap the wood underneath the blade and against the fence, so it will have nowhere to go if it binds except back at you.

Ripping bevels is easy: tilt the blade to the angle you need, and set the rip fence the correct distance from the blade for your workpiece. It's a bit more difficult to measure off of a tilted blade to set the rip fence than if the blade is square to the table. To be sure your rip fence is set correctly, start the saw, and slide your workpiece up to the blade until the teeth just nick the workpiece. Back the board away from the blade and check to see that the nick aligns accurately with the cutting mark you've drawn. Adjust the fence as necessary and proceed with the cut.

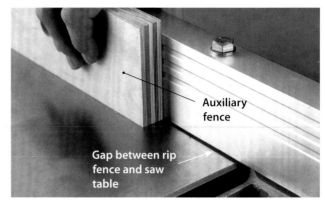

Inspect the joint between the rip fence and saw table with the fence locked down. Before you cut thin stock, like plastic laminate, cover any gaps by bolting an auxiliary fence to your rip fence.

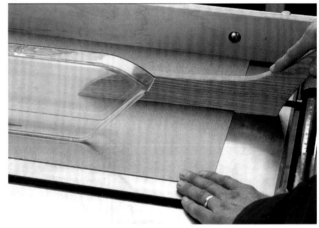

Hold laminate or other thin sheet materials firmly against the saw table with a long-soled push stick to keep the workpiece from lifting or fluttering against the blade as you cut.

Set up beveled rip cuts so the blade tilts away from the rip fence. This keeps the workpiece from becoming trapped between the fence and the blade. Clamp featherboards to the saw to hold down larger panels.

Bevel-Cutting on Edge

From time to time, you'll need to cut bevels at angles exceeding the maximum blade tilt of your saw. In these instances, you'll need to stand the workpiece on edge against the rip fence to cut the bevel. For workpieces taller or wider than about 6", attach a tall wooden auxiliary fence to the rip fence to support the workpiece as you cut. Without a tall fence, it's tough to keep a board from teetering away from the fence, which could bind on the blade and lead to kickback.

Any flat, wide scrap of ¾"-thick stock will make a suitable tall fence, but it should extend the full length of your rip fence. A tall fence must sit perpendicular to the saw table and should not distort when you attach it to your rip fence. You can build a tall fence that clamps to the saw's rip fence (see pages 154–155) or bolt one directly to the rip fence. However it attaches, the tall fence should be slightly shorter than the workpiece you are cutting so that you can keep a firm grip on the workpiece as you cut. If the beveled edges of the workpiece

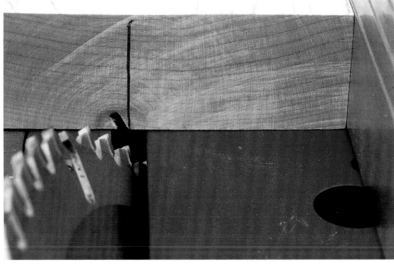

Feed the workpiece into the blade until the blade just nicks the board. Use the nick as a guide for adjusting the rip fence until the blade aligns accurately with your bevel layout mark.

are narrow, clamp the workpiece to a piece of scrap that rides on top of the rip fence to keep the workpiece from diving into the blade slot opening in the throat plate. Or, install a zero-clearance throat plate (see pages 41–42).

Cutting Raised Panels on the Table Saw

Magnetic featherboard

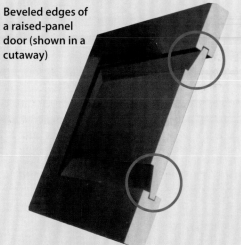

Beveled edges of a raised-panel door (shown in a cutaway)

Raised panels form the center portion of many traditional cabinet and full-sized doors. One face of the panel is beveled on the edges and ends so it can fit in a groove in the surrounding doorframe (see the cutaway drawing, above). To cut the panel, you'll need to bevel-rip the workpiece on edge against a tall fence. Clamp on a featherboard to hold the workpiece tight against the fence. Install a zero-clearance throat plate in the saw, or clamp a piece of scrap to the workpiece to ride along the top of the fence.

Ripping Tapers

Tapers are modified rip cuts made with a jig that holds the workpiece at an angle to the saw blade. The workpiece starts out square on all faces and edges, but the jig slides along the rip fence and allows the blade to cut into the workpiece at an angle, so one end of the workpiece becomes narrower than the other.

In furniture building, tapering is a fairly common technique for slimming down the profiles of table and chair legs. Tapers are easy cuts to make, but there is no safe way to cut a taper on the table saw without using a jig. Never attempt to cut a taper by setting the rip fence at an angle to the blade or pushing the workpiece freehand through the blade. The wood will kick back.

Tapering Jig Styles

Tapering jigs can be either fixed or adjustable. Unless you cut many tapers, a simple fixed tapering jig is probably all you need. It's simply a rectangular piece of scrap stock with a cutout area along one edge that matches the taper profile. Follow the directions on the next page to build a fixed tapering jig.

The adjustable variety is just as easy to use as a fixed jig, but it takes a little longer to build. (For instructions on building an adjustable tapering jig, see pages 156–157.) You can also buy relatively inexpensive adjustable tapering jigs through most woodworking stores and catalogs.

Adjustable tapering jigs consist of two "arms" about 2' long that are attached together on one end with a hinge. One arm rides against the rip fence, while the other arm supports the workpiece. A stop at the end of the support arm holds the workpiece in position as you slide the jig and workpiece through the blade. An adjustable brace spans the two support arms on top of the jig and holds them at a fixed angle that matches the taper on the workpiece.

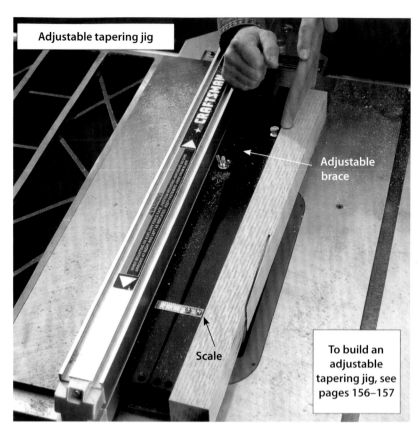

Adjustable tapering jig

Adjustable brace

Scale

To build an adjustable tapering jig, see pages 156–157

Adjustable tapering jigs can be set to cut tapers of various degrees of shape. Most jigs are made of metal and come with an adjustable brace that spans the "arms" of the jig to lock them in position. Some have a scale for setting taper angles.

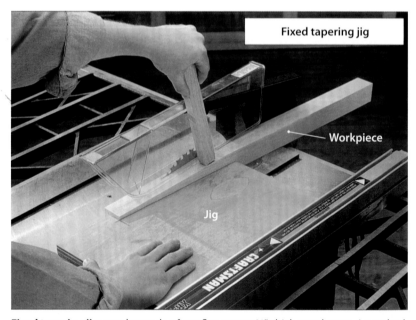

Fixed tapering jig

Workpiece

Jig

Fixed tapering jigs can be made of any flat, square ¾"-thick panel scrap. A notched cutout area along the edge of the jig establishes the taper angle. To cut a taper, set the workpiece into the notch on the jig and slide the jig along the rip fence. Use a push stick or scrap to hold the workpiece against the saw table and the jig.

Building a Fixed Tapering Jig

Fixed tapering jigs begin with a rectangular piece of ¾" plywood. The size of the jig depends on the length of the taper you plan to cut. The longer the taper is, the longer the jig will need to be, because the jig marks the full length of the material the saw will remove to form the taper. The jig should be wide enough so there is plenty of material between the tapering area and the flat edge that will ride against the rip fence of your saw.

Using the upper left corner of the plywood as a starting point, measure in along the top edge the thickness of the tapered workpiece at its thickest point. Then measure down the edge of the jig and mark the length of the taper. Measure in straight from that point and mark the thickness of the taper at its narrow end. Connect the points you just made with a straightedge, and cut out the notched area along the outline with a jigsaw or band saw. The area you have removed forms the taper profile.

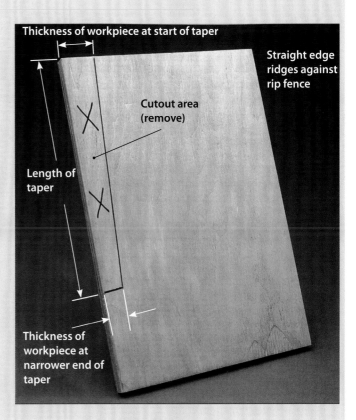

Thickness of workpiece at start of taper

Straight edge ridges against rip fence

Cutout area (remove)

Length of taper

Thickness of workpiece at narrower end of taper

Fixed Tapering Jigs

To use a fixed tapering jig, set the jig on the saw table with the long, flat edge resting against the rip fence. Set the rip fence so the distance between the blade and fence matches the full width of the jig; the blade should only cut into the workpiece, not the jig. Then raise the blade high enough to cut through the workpiece in one pass. Set the workpiece in the notch and back the jig up on the saw table so the workpiece is clear of the blade. Grab a push stick to hold the workpiece tight against the jig with your left hand as you push the jig with your right hand. Start the saw and slide the jig forward slowly with your right hand until the blade engages the workpiece. Then feed the jig steadily along the rip fence, keeping the workpiece held snugly against the jig with a push stick until the jig clears the blade.

If you want to taper two adjacent faces of the workpiece, flip the tapered edge up, set the workpiece in the jig, and make another pass to cut the second taper. In cases where you want to cut a taper on each face of a workpiece, you'll need to temporarily glue the wedges you cut free on the first two tapers back into place to create bearing surfaces for cutting tapers on the third and fourth faces of the workpiece. Hot-melt glue works well for this purpose. Remove the wedges afterward.

Table Saw History

Swing-Blade Off Cut Saw

Built to be semi-portable, this cutoff saw was unique because the fence was fixed permanently to the saw table. Rather than sliding workpieces over the table, the operator would step down on a foot pedal in front of the saw to swing the arbor and blade across

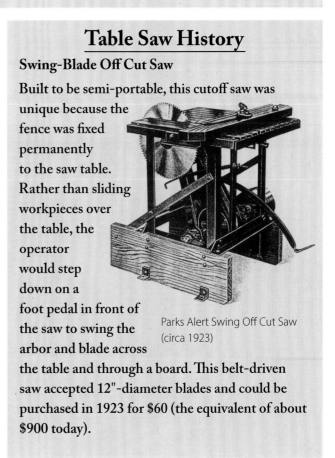

Parks Alert Swing Off Cut Saw (circa 1923)

the table and through a board. This belt-driven saw accepted 12"-diameter blades and could be purchased in 1923 for $60 (the equivalent of about $900 today).

Setting Up a Taper Cut

To set up for cutting tapers with an adjustable tapering jig, first mark layout lines on the workpiece to indicate where the taper starts and ends. Raise the blade a couple of inches above the saw table.

1

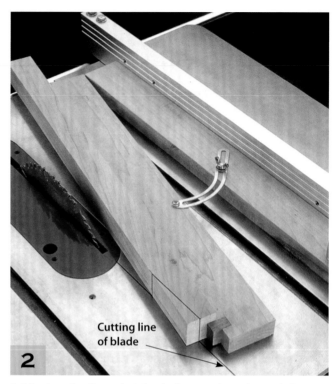

2

Cutting line
of blade

Lay a long straightedge flat on the saw table and against the side of the blade that faces the rip fence. Draw a line across the saw table to mark the cutting plane of the blade (see photo, above). This will serve as a reference line to align the tapering jig. Then remove the straightedge and lower the blade below the saw table.

Set the tapering jig against the rip fence and rest the workpiece against the jig. As a unit, slide the rip fence and jig toward the blade, aligning the layout lines on the workpiece with the blade cutting line on the saw table. Lock the rip fence in position. Then lock the tapering jig to this cutting angle by tightening the jig's cross brace.

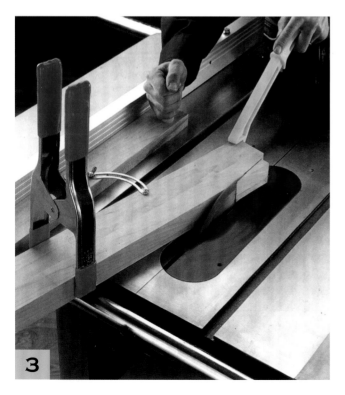

3

Back the jig and workpiece up so they clear the blade, and raise the blade high enough to cut through the workpiece in one pass. Slide the jig along the rip fence to cut the taper. Use a push stick or spring clamp to hold the workpiece against the jig as you make the cut. If you need to cut a taper on an adjacent face, the jig is already set to go. Flip the workpiece so the tapered edge faces up and cut a second taper. To taper all four faces, double the cutting angle of the jig when the tapered faces rest against it.

MAKING CROSSCUTS

Next to rip cuts, crosscuts are the second most common cuts you'll make on a table saw. Crosscutting is just what its name implies—cutting a workpiece across the grain, which in most cases means cutting a board to length. The basic steps involved in crosscutting on a table saw are setting one long edge of a board against the miter gauge fence, aligning the layout line on the board with the blade, then feeding the board through the blade.

Typically, crosscuts are made after a workpiece has been ripped to width. One crosscut is made to square the end of the board, then a second crosscut saws the board to length. You'll rarely perform rip cuts on a workpiece without then following up with a couple of crosscuts.

Safer Cuts

Crosscuts are considerably safer to make than rip cuts for one main reason: You'll use the miter gauge, not the rip fence, as your guide. Remember that ripping involves sandwiching a workpiece in between the rip fence and the blade, where it can bind, get caught on the saw teeth at the back of the blade, and kick back. Crosscutting, on the other hand, doesn't restrict the clear space around the workpiece. In fact, the rip fence shouldn't even touch the workpiece during a crosscut while the blade is engaged in the wood.

One kind of kickback can occur when crosscutting, but it is the waste piece—not the workpiece—that poses the hazard: Unlike the strips of waste produced by rip cuts, which typically are long enough to be kept clear of the blade by the splitter, crosscutting often creates short pieces of waste that sit right next to the blade when they are cut free. If you don't use a zero-clearance throat plate on the saw, small waste pieces can get pulled down into the gap between the blade and the throat plate. If this happens near the back of the blade, the teeth on the outfeed side of the blade could lift the waste and throw it back in your direction. A blade guard and a zero-clearance throat plate will significantly reduce the chances of injury from this kind of kickback.

> ## Caution
>
> Some of the photographs you'll see throughout this book show the blade guard and splitter or riving knife removed. This was done to make the information in the photographs easier to see. To prevent injury, use a working blade guard and a splitter or riving knife with anti-kickback pawls whenever the type of cut being made allows it.

Crosscutting, or sawing wood across the grain, serves two purposes: it squares the ends of workpieces and trims them to length. You'll perform these cuts with workpieces held against the miter gauge.

Table Saw History

Revolving Twin-Arbor Saw

Built to meet the sawing needs of cabinet, pattern, sash, and carriage shops in the early 1900s, this saw featured a double arbor cradle that pivoted on a center shaft so the saw could be outfitted with two blades—one for crosscutting and the other for ripping. A crank below the saw table revolved either type of 14"-diameter blade into position. The saw table tilted left and right up to 45° for making bevel cuts. The right side of the cast-iron table was fixed in position for making rip cuts, but the left side, outfitted with twin mitering fences, actually slid past the blade on rollers for making crosscuts.

J. A. Fay & Co. Double Circular Sawing Machine (circa 1903)

Crosscutting with a Miter Gauge

Most crosscuts are intended to square the ends of workpieces. To make these cuts accurately, first set the miter fence perpendicular to the miter bar with a square. While you're at it, use a square to make sure that the saw blade is perpendicular to the saw table. Once the blade is square to the table, lower it to a height suitable for cutting through the workpiece. Be sure the splitter is aligned with the blade.

Setting Up the Cut

It doesn't matter which miter slot you choose for making square crosscuts with the miter gauge. If one of the two miter slots on your saw is milled closer to the blade than the other, choose the closer slot for crosscutting, so the miter fence will support the workpiece as close to the blade as possible. The closer you can support the workpiece to the blade, the less likely it will be to shift away from the blade as you cut.

Draw layout lines across both the face and leading edge of the board (the edge that will make contact with the blade first). With the saw turned off, set one flat edge of the workpiece against the miter fence and slide the miter gauge forward on the saw table until the cutting line on the edge of the board touches the blade. Align the layout lines

with the blade so that the blade will cut on the waste side of the workpiece.

Once you've aligned the board with the blade, clamp or hold the workpiece in position against the miter fence. Stand behind the miter gauge so you are clear of the path of the blade. If you hold, rather than clamp, the workpiece to the miter gauge, wrap your hand around the fence to grip the work and push the gauge forward with your other hand on the miter fence lock knob. This way, your arms should not cross the blade at any time during a cut.

Making the Cut

Start the saw, slide the miter gauge forward slowly until the blade enters the wood, then increase your feed rate to cut quickly and cleanly through the workpiece without stopping. Continue pushing the miter gauge forward until it completely passes the blade.

As soon as the full workpiece passes the infeed side of the blade, the waste piece will be cut free and stop next to the spinning blade. Do not let waste pieces accumulate here. If the waste is just a sliver, the blade will probably blow it out of the way or suck it down the blade slot opening in the throat plate. Larger cutoffs will sit on the table or may even rattle against the blade. It's important to clear these waste pieces away from the blade to keep them from accidentally catching on the blade, but only if they are large enough to be removed safely while the blade is spinning. Use a push stick to slide cutoffs shorter than 6" away from the blade. If you can't remove waste pieces easily with a push stick, shut off the saw and brush them away once the blade has stopped.

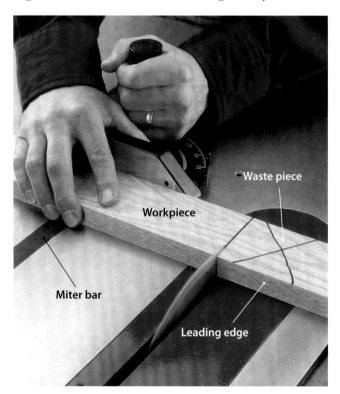

When making crosscuts, set the longer section of the workpiece against the miter gauge and cut the shorter portion free. Usually, the part you hold against the miter gauge will be your workpiece rather than a waste piece. Align the blade so it cuts on the waste side, rather than the workpiece side, of the board.

If you are crosscutting a panel that is too wide to be fed into the blade with the miter gauge in back of the board, position the gauge ahead of the panel instead. Attach an auxiliary fence to the miter gauge to provide more support. Feed the panel across the blade as usual, from the infeed side of the saw.

Improving Miter Gauge Performance

Although a miter gauge makes crosscutting safer, it doesn't necessarily make crosscuts precisely. Most miter gauges come with fences that are only about 6" long, sometimes even less. The trouble is, most workpieces you'll need to crosscut will be longer than the miter gauge fence. The longer the workpiece, the harder time you'll have holding it tight against the short miter fence as you feed it through the blade. This is because the center of gravity of a board is near the middle, which moves farther and farther away from the miter gauge the closer to the end of the board you cut (see illustration, below). The weight of the board acts like a lever against the short portion you are clamping or holding against the miter fence. The result? Long workpieces tend to pivot away from the miter gauge fence toward their center of gravity. The instant this happens, your cut will shift away from the layout line, ruining your accuracy.

The problem here boils down to poor workpiece support, but you can solve miter gauge limitations in three fairly easy ways: add an auxiliary fence to the miter gauge; build a shooting board; or build a crosscut sled.

Adding an Auxiliary Miter Fence

Most miter gauge fence bodies come with two holes drilled through the fence for adding an auxiliary wooden fence. The auxiliary fence will get chewed up by the blade, so it's an easy item to periodically replace. To make one, select a piece of dimensionally stable scrap wood (hardwood, MDF, or plywood) about 2' long, ¾" to 1" thick, and about 1" wider than the saw's full blade height above the table. The auxiliary fence should be flat on both faces and square on at least one long edge.

Squaring Board Ends

When squaring up the end of a board, make the cut far enough in from the end so there is a significant amount of wood on either side of the blade. Don't shave board ends. A table saw is a cutting tool—not a shaving tool. If cutting stresses aren't distributed evenly on both sides of the blade, especially on thin-kerf blades, the blade will deflect. Once the blade deflects, your cut won't be square. Check the cut end with a square to verify that it is actually square to the board's edges and faces.

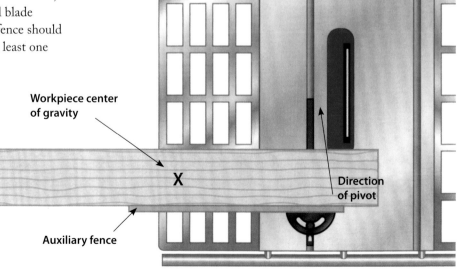

Workpiece center of gravity

X

Direction of pivot

Auxiliary fence

The closer you cut to the end of a workpiece, the farther away its center of gravity moves from the miter gauge. Without an auxiliary fence attached to the miter gauge, the workpiece will pivot at its center of gravity like a lever, away from the miter fence.

You have two choices for building auxiliary miter fences—flush-style or sweep-style. Flush-style auxiliary fences stop just short of the saw guard, so the guard doesn't have to ride up and over the fence when you make a cut.

Sweep-style miter fences extend about 6" or so past the blade, so they provide better workpiece support around the blade than flush-style fences. Plus, the added fence length beyond the blade will sweep waste pieces clear when you push the miter gauge through the cut.

Since the saw kerf on a sweep-style fence matches the blade kerf exactly, use the kerf on the fence to align the cut lines on your workpieces with the blade. Be aware, however, that the guard on your saw will have to travel up and over a sweep-style fence, because it straddles the blade. If the guard on your saw cannot rise up and over the sweep fence, attach a flush-style fence instead. Do not solve this problem by removing your saw's guard and splitter or riving knife.

Adding an auxiliary fence. Most miter gauge fences are outfitted with a pair of holes to accommodate screws for attaching an auxiliary fence. Select a flat piece of scrap wood or plywood for the auxiliary fence. Set the miter gauge in a miter gauge slot on the saw table. Attach the fence with short wood screws so it sits flush with the saw table. Screws should be short enough so they don't pierce the front of the fence.

Tip

To attach either fence style, choose screws short enough so the tips do not pierce the front face of the auxiliary fence. It's helpful to cut a piece of medium-grit sandpaper sized to cover the fence face and use spray-mount adhesive or glue to adhere it to the fence, grit side facing out. Sandpaper will give the fence "tooth" for a better grip on workpieces.

Common Auxiliary Miter Fence Styles

Fence stops short of saw guard and blade

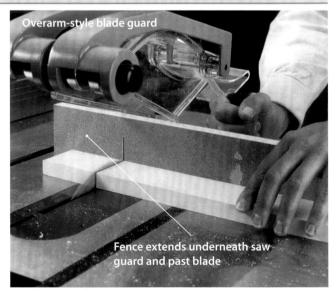

Overarm-style blade guard

Fence extends underneath saw guard and past blade

Flush-style miter fences stop short of the blade guard to keep the guard from riding up and over the fence. This style also keeps the fence clear of the posts that mount the guard to the splitter or riving knife, which would obstruct the path of the miter gauge.

Sweep-style miter fences extend past the blade and provide a sweeping effect for waste pieces, once they are cut free. Your saw guard will have to ride up and over the fence, so this fence style is best suited for overarm-style blade guards.

Shooting Boards

Workpieces longer than 2' are tricky to crosscut accurately on a table saw. When workpieces extend past the ends of your auxiliary miter fence, they become progressively harder to manage. Depending upon how long your workpiece is and where along its length you plan to make your cut, it will be tough to hold the workpiece steady against even a long auxiliary miter fence. One solution to this problem is to add a support device alongside the saw, but you'll still have to drag the workpiece across the saw table to cut it when guiding it with the miter gauge. Another option is to build a jig that supports the workpiece from beneath, called a *shooting board*. (Shooting boards are also commonly known as *panel-cutting jigs*.)

A shooting board can replace the miter gauge for crosscuts on longer or wider workpieces. It consists of a piece of plywood with a hardwood runner that slides into one of the two miter slots on the saw table (see illustration, below). The jig's plywood base can be as large as half the saw table, including an extension wing. The bigger, the better. Shooting boards are equipped with a long fence along the trailing edge of the jig base to provide a surface that holds the workpiece. The fence also allows you to spread your arms apart when you push the jig, distributing the pushing action more evenly between both arms.

Shooting board

Shooting boards provide a more stable support system for cutting long workpieces because the workpiece actually rides on top of the jig rather than sliding on the saw table. A long fence on the back of the jig keeps larger workpieces square with the blade. The jig rides on a hardwood runner that fits in the miter slot.

Building a Shooting Board

Cut the base of the shooting board from a piece of ¾" cabinet-grade plywood, melamine-covered particleboard, or MDF. Size the base so it covers the full depth of your saw table and is a couple inches longer than the distance from the blade to the outer edge of one extension wing. Cut the miter slot runner from hardwood. The runner should slide into the miter slot without binding and be flush with the surface of the saw table. Slip the runner into the miter slot. The shooting board base piece should be positioned so one end of the base extends past the blade about an inch. Square the base to the front edge of the saw table. Mark the position of the runner on the base, and attach the runner to the base with short countersunk wood screws. Set the base and runner in the miter slot again and trim the end of the jig base so it is flush with the blade. Cut the fence from a piece of 2"- to 3"-wide hardwood. The fence length should match the length of the base. Use a carpenter's square against the "blade" end of the base to square the fence along the front edge of the base. Attach the fence to the base with 1¼" countersunk wood screws.

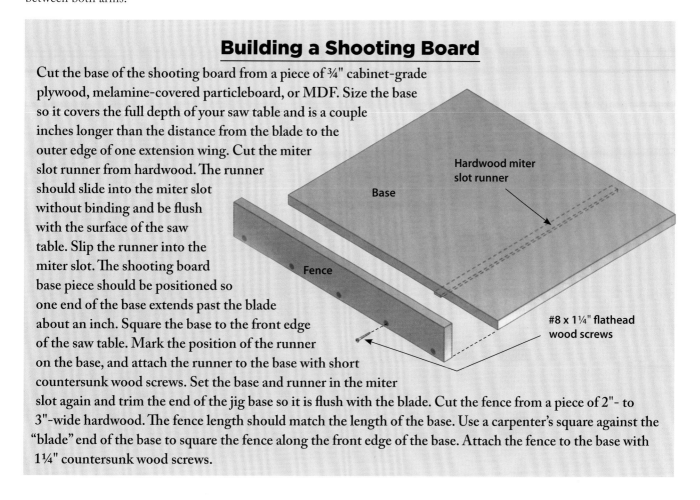

Base

Hardwood miter slot runner

Fence

#8 x 1¼" flathead wood screws

A shooting board outperforms a miter gauge for making crosscuts because it carries a workpiece across the saw table rather than sliding it on the table as a miter gauge does. Dragging the workpiece over the saw table is the main reason why workpieces shift during typical miter gauge cuts.

When you construct a shooting board, you'll trim the jig so one end of the base and the fence are flush with the blade (see page 89). With this arrangement, aligning crosscuts is easy. Simply set the workpiece on the shooting board so the cutting line and the edge of the fence match up. This way the blade will cut up to and on the waste side of your cutting line.

Making Crosscuts with a Shooting Board

Before you start the saw, be sure the blade guard will rise up and over the vertical fence on the shooting board. Back the shooting board up on the saw table so the shooting board is clear of the blade. Stand out of line with the blade behind the shooting board. Start the saw and slide the board forward until the blade engages the workpiece, then slide the shooting board and workpiece smoothly past the blade.

The waste piece will drop next to the blade as soon you finish the cut. Unless the waste piece is at least 6" long, avoid the temptation to slide it away from the blade with your hand or flick it away with a push stick. The safest method for clearing away a short cutoff scrap is to shut down the saw, wait until the blade stops, and slide the waste away from the blade with a push stick before you pick it up.

Aligning a workpiece on a shooting board. It's easy to precisely align a workpiece cutting line with the blade on a shooting board, because the edge of the jig is flush with the side of the blade. Align the workpiece cutting line with the edge of the shooting board base, hold the workpiece against the fence, and make your cut. You'll need to raise the blade high enough to account for the thickness of the jig base in order to cut through the workpiece in one pass.

Table Saw History
Cut Off Saw Manufactured to Build Railways

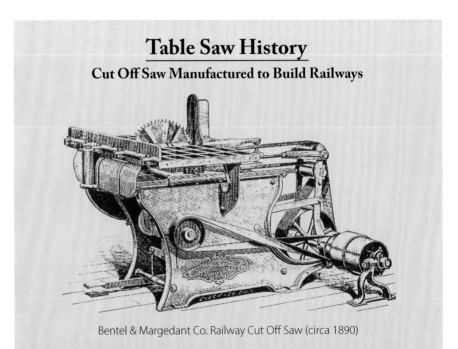

Bentel & Margedant Co. Railway Cut Off Saw (circa 1890)

Built in the 1890s to meet the sawing needs of America's growing railroad system, this massive 1250-lb. machine likely crosscut timbers used for railroad ties. The saw featured a sliding tabletop for feeding timbers into the blade. The main drive belt could be adjusted on a cone-shaped pulley to change blade speed.

Crosscut Sleds

Crosscutting with a crosscut sled is like using a pair of shooting boards in tandem on either side of the blade. Unlike a shooting board, which only supports a workpiece on one side of the blade, a crosscut sled cradles both the workpiece and the waste piece.

The sled has a multitude of uses, and you'll see different examples of its versatility throughout this book, but it shines as the superior crosscutting jig. You can use it in place of the miter gauge for typical crosscuts or even for making miter cuts if you add a simple mitering attachment (see page 96). A crosscut sled is especially handy when you need to crosscut long workpieces that will produce long, unwieldy cutoffs. When you use a shooting board for the same task, long cutoffs will drop next to the blade as soon as they are cut free, where they could accidentally come into contact with the blade or even tip off the saw table. This won't happen on a crosscut sled. Even when the waste piece is cut free, it still rests on the sled base.

Crosscut sleds provide support for workpieces on both sides of the blade, which is especially helpful for crosscutting long pieces of lumber into two long sections. After you make the cut, both lengths of stock are still supported from beneath. Clamp a scrap block against the sled fence to keep workpieces held snugly against the sled base.

Crosscut Sled Features

Most crosscut sleds have basically the same design. A crosscut sled starts out as a piece of plywood about the size of the full saw table plus extension wings. It has two shooting board–style runners fastened below to ride in the miter slots. A tall fence spans either long edge of the sled. Once the jig is made, you'll raise the saw blade and cut a kerf across the entire width of the sled and through the fences to create a blade track. See pages 151–153 for a complete set of plans to build the crosscut sled you'll see featured in this book.

The fence on the infeed side of the sled serves the same purpose as the fence on the shooting board—to support, align, and anchor workpieces. The fence on the outfeed side of the jig bridges the gap created by the saw kerf and holds the two plywood base pieces in place. Since the sled covers the full saw table, the jig fences will interfere with most saw guard and splitter assemblies. Unless your guard mounts from the extreme left-or right-hand side of the saw, you'll have to remove these safety items in order to use a crosscut sled.

Some crosscut sled designs incorporate safety features to shield you from blade exposure. The crosscut sled plans featured in this book include information for constructing a simple blade guard and blade tunnel. If the plans you build from do not call for a blade guard, add one. Without a blade guard, a crosscut sled does nothing to protect you from blade injuries.

Using a Sled

To use the sled, set your workpiece against the infeed fence, aligning the cutting line on the workpiece with the saw kerf in the fence. Clamp or hold the workpiece tightly to the fence. If your workpiece is narrow, you may be able to start the saw with the blade inside the sled. This is perfectly safe, provided the blade guard is in position and there is plenty of space between the front edge of the workpiece and the blade. For wide panels, you'll have to start the cut with the crosscut sled pulled in front of the blade.

Stand to the right or left of the blade, behind the side of the sled that supports the heaviest part of what you are cutting. Slide the sled across the table until the blade cuts through the board. Turn off the saw and remove both cut pieces. If you are making repetitive cuts and do not want to shut the saw down between cuts, slide the workpiece and waste piece away from the saw kerf in the sled and remove the waste piece. Then pull the sled back toward yourself to set up for the next cut. Do not let waste pieces accumulate in the sled, where they could accidentally come in contact with the blade.

Crosscutting Long Boards

Side Support Options

Sawhorse extension. Adding side support doesn't have to be fancy—just stable. One easy option is to screw a couple of short legs to a length of scrap stock, then clamp the assembly to the top of a sawhorse at whatever saw table height you need.

Outfeed table. Build the outfeed table shown on pages 160–161 and use it as a side support device as well as an outfeed table. This accessory clamps into the vise of a portable workbench, so its height is fully adjustable. If you don't own a portable workbench, you could also clamp the jig to a sawhorse.

Tip

The techniques you use for making safe, accurate crosscuts will need to change somewhat to accommodate workpieces of different proportions. You'll also want to modify your crosscutting approach when you pivot the miter gauge to make mitered cuts. The next few pages cover a number of specialized crosscutting situations and procedures you should follow for making these cuts.

Crosscutting a 6' or 8' board is cumbersome and tough to do with great precision using just the miter gauge as a guide. Longer boards will become even more unwieldy on the table saw. Power miter saws are gaining popularity by both contractors and woodworkers because these tools allow you to hold a long board stationary and pull the saw blade through them, rather than the other way around as you must do on the table saw. For extremely long stock, choose the most suitable saw for the job. In some cases, a power miter saw, circular saw, or even a crosscut handsaw may make more sense than maneuvering a long board over a short saw table. You can always cut long stock into shorter segments with one of these saws, then make final crosscuts on the table saw.

Reinforcing a Miter Fence

Stiffen long auxiliary miter fences to keep them from deflecting by attaching a wood spine perpendicular to the back of the fence. The spine should be narrow enough so it doesn't obstruct the miter gauge handle.

When your only option is to crosscut a long board on the table saw, use a shooting board or crosscut sled to support as much of the board on the saw table as you can, or fasten a long auxiliary fence to the miter gauge. Strengthen the auxiliary miter fence with a wood spine to keep it from deflecting (see tip, above). If a workpiece extends more than a couple feet over the edge of the saw table, set up a tall sawhorse or another worktable next to the saw to support the workpiece as you make your cut.

Crosscutting Short Stock

Whenever you can, crosscut a workpiece from a piece of stock long enough to handle safely. It's not always practical to do this, however, if all you have to work with is a short piece of stock. In situations where your hand would need to come within 6" of the blade to hold a workpiece for a crosscut, make the cut on a shooting board or crosscut sled instead of using only the miter gauge. A short workpiece is more likely to shift when dragged over the saw table against a miter fence, especially if you have to hold it down with a push stick as you feed the workpiece into the blade. Hold the short piece in place on the shooting board or crosscut sled with a rubber-soled push pad, a piece of scrap, or a push stick.

Making Repetitive Crosscuts

Crosscutting short stock. The safest method for crosscutting short stock is to make the cut on a crosscut sled. Keep your hand a safe distance from the blade by holding the workpiece with a rubber-soled push pad, a piece of scrap wood, or a push stick. If you don't have a crosscut sled, crosscut short workpieces from longer lengths of stock—never risk your fingers by holding workpieces shorter than 6" by hand.

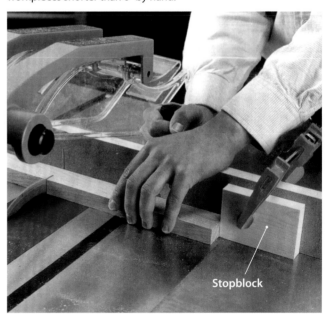

Making repetitive crosscuts with a miter fence. Clamp a stopblock to an auxiliary miter fence and use the block as a guide for crosscutting many workpieces to the same length. The distance from the near side of the blade to the stopblock equals the length of your workpiece, once it's crosscut to size.

Stopblock

Making repetitive crosscuts with a rip fence. Clamp a stopblock to the rip fence on the infeed side of the blade. Clamp the block far enough back from the blade so your workpiece will clear the stopblock before the blade engages it. This setup will provide a clear space between the rip fence and the end of the workpiece. The distance from the stopblock to the blade equals the length of the workpieces, once they are cut to size.

Frequently you'll need to crosscut a number of workpieces to the same length for a project. You could measure these parts out one by one and crosscut each part to length, but this method is time-consuming. Plus, parts cut to length this way seldom match.

A faster and more accurate method for making repetitive crosscuts on a table saw is to use a stopblock. A stopblock is simply a piece of scrap with square faces and edges that's large enough to be clamped in place on the saw. The common places to clamp a stopblock are to the fence, to an auxiliary fence on a miter gauge, or to the fence on a shooting board or crosscut sled. You'll clamp the stopblock a precise distance away from the blade, then position one end of the workpiece against the stopblock, using the stopblock as an index for aligning the blade with your cutting line. This way, you only have to measure once to establish a fixed length for cutting as many pieces as you need.

When using a stopblock attached to an auxiliary miter gauge fence or the fence on the crosscut sled or shooting board, clamp the stopblock to the fence on the same side of the blade as the workpiece. The distance from the stopblock to the blade should match the finished length of the workpiece. If the workpiece length exceeds the length of the auxiliary fence, install a longer fence.

As a guide for making repetitive crosscuts, you can also clamp a stopblock to the rip fence. The stopblock should be on the infeed side of the blade, near the front of the saw table. Be sure the stopblock is at least ¾" thick. This cutting

situation is the only time it is safe to use the miter gauge and the rip fence in tandem when cutting a board into parts. The distance from the stopblock to the blade sets the length of the workpieces. The stopblock also provides a space between the end of the workpiece and the rip fence. Without this space, the cutoff piece would be trapped between the rip fence, miter fence, and the blade, which could result in kickback.

To set up a stopblock on the rip fence, clamp the stopblock far enough forward on the infeed side of the rip fence so the workpiece will clear the stopblock before it makes contact with the blade. Set the workpiece against the miter gauge and slide it along the miter fence until the cutting line aligns with the blade. Hold the workpiece in this position and pull the miter gauge back to the front of the saw table. Slide the rip fence over until the stopblock touches the end of the workpiece, then lock the rip fence in position. Before you make each cut, be sure the workpiece touches the stopblock.

Crosscutting Bevels

The setup for beveled crosscuts is nearly the same as for square crosscuts, except the blade is set at an angle other than 90°. When setting up a beveled crosscut, it's important to use the miter slot that faces away from the direction of blade tilt. For instance, if the blade tilts to the right, make the cut with the miter gauge on the left side of the blade. The objective here is to cut the workpiece so the waste is below the blade, not above it. If the waste piece ends up on top of the blade once it's cut free, it could catch on the blade and kick back.

Making beveled crosscuts. Set up beveled crosscuts so the blade tilts away from the miter gauge. This way, waste pieces will rest against the saw table once they are cut free rather than on top of the blade, where they could catch on the blade teeth and shoot back.

MAKING MITER CUTS

Cutting miters involves swiveling the protractor-style miter gauge head so it is no longer perpendicular to the blade. The most common miter setting is 45°, but miter gauges can be adjusted to any angle up to 45°. Unless you are using a precision miter gauge (see page 143), don't trust that the scale on your miter gauge is accurate. For maximum accuracy, use a drafting triangle or bevel gauge set against the blade and miter gauge to establish exact angles. If you want to match an angle already scribed on your workpiece, lay the miter gauge head against a flat edge of the workpiece and swivel the miter bar until it matches the angle on the workpiece. Once you've determined the miter angle and set the protractor head, it never hurts to make a test cut on a scrap piece before cutting your workpiece.

The only potentially frustrating part of making miter cuts is that it's tougher to hold a workpiece steady against the miter gauge with the head at an angle than when the head is square to the blade. Boards will want to creep away from the blade along the miter fence in response to forces imposed by the blade and friction from the saw table. The steeper the miter angle setting, the worse your control will become. It's tough to make accurate miters without first attaching an auxiliary fence to the miter gauge.

Making miter cuts. Attach an auxiliary fence to the miter gauge to support your workpiece right up to the blade. For best control when miter-cutting, set up your cuts so the cut end leads the rest of the workpiece across the table, as shown above.

Methods for Setting a Miter Gauge

Use the workpiece cutting line. When the angle you need to cut is already marked on your workpiece, match the angle on the miter gauge by setting the bar along your layout line and the protractor head flush against one edge of the workpiece. With this method, you'll have to flip the workpiece layout-side down on the saw table to orient it properly on the miter gauge, so mark both workpiece faces.

Use a bevel gauge to set angles. In cases where you want to cut a miter to match an odd angle taken from another source, use a bevel gauge to mark the angle you need. Then set the bevel gauge base against the saw blade and swivel the miter gauge fence until it is flush with the bevel gauge arm. Lock the miter gauge at this angle.

When you can, it's best to orient the workpiece on the miter gauge so the cut will lead the workpiece past the blade, rather than follow behind it. Pushing the cut ahead of the rest of the workpiece will keep you from reaching farther than you need to over the saw. You'll also have better hand control over the workpiece.

Attach a stopblock to the miter fence if you need to make repetitive miter cuts, just as you would when making repetitive square crosscuts. If a stopblock needs to support an end that's already mitered, it must stand out far enough from the fence to make contact with the tip of the miter. When you can, set the mitered end of a workpiece so it tapers toward the fence, rather than away from it. Clamp an angled stopblock against the workpiece to capture the mitered end of the workpiece against the fence (see photo, right).

Mitering Jig for Crosscut Sleds

Turn your crosscut into a first-rate mitering accessory by building the simple miter jig shown here. We used a 14"-square scrap of tempered hardboard for the base (A), cut diagonally into the shape of a 45° triangle. Attach a fence (B) along the long edge of the hardboard and a couple of scrap blocks (C) behind the fence on each end. The blocks serve as clamping points for holding the miter jig in place on the sled. To use the mitering attachment, clamp the square corner of the jig base against the sled fence so one short edge of the jig is flush with the saw kerf in the sled.

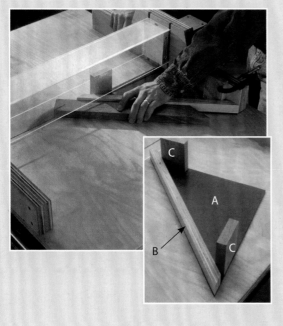

Making repetitive miter cuts with a stopblock. You can make repetitive miter cuts by clamping a stopblock to the miter gauge fence, just as you would if you were making repetitive crosscuts. If one end of each workpiece is already cut at an angle, capture the mitered end against the auxiliary miter fence with a stopblock cut at a complementary angle.

Compound miter cuts are relatively common cuts you'll need to make for projects such as installing cove and crown molding. To make a compound miter cut, you'll tilt the blade and swivel the miter gauge head to cut two angles at once. Test your setup on a scrap piece before committing to your actual workpiece.

CUTTING SHEET GOODS SAFELY

From time to time, your woodworking or home improvement projects will require you to use manufactured sheet goods, such as MDF, hardboard, particleboard, or various kinds of plywood. These products have several advantages over board lumber: they don't expand or contract to the same degree as wood, and they have no grain direction, so they aren't prone to twisting and warping like actual lumber. Plus, sheet goods typically come in 4' x 8' panels, which enable you to create large panels without having to glue a number of boards together.

In spite of their advantages as a building material, full-sized sheets of plywood or particleboard are heavy to lift and can be unwieldy to control, especially if you are working alone. Weight and workpiece proportions are the main issues you'll have to contend with when sawing sheet goods on a table saw.

Pointers for Safer, Smoother Cuts

Sheet stock can be broken down safely and accurately on the table saw, but here are a few rules of thumb to keep in mind:

1. Start with Manageable Workpieces

Many woodworkers cut 4 x 8 sheets into smaller pieces with a circular saw or jigsaw first to make them more manageable on the table saw. This is an especially good practice to follow if you are using a jobsite saw with a small table, or if you don't have a number of sturdy worktables, roller stands, or sawhorses handy to support your work. Lay out a piece of sheet stock so you'll make the longest rip cuts and crosscuts first. This way, you'll first break the sheets down into reasonable pieces. Then subdivide the smaller pieces into individual parts on the table saw.

2. Guards and Splitters/Riving Knives are a Must

Larger sheet panels will require you to start your cuts farther away from the saw, which makes it harder to keep a large workpiece tight to the rip fence. The problem compounds if you are also straining to support a heavy panel or if the panel is thin and starts to bow. If the workpiece drifts away from the rip fence, you are at a greater risk for kickback, so always saw sheet goods with a guard and splitter or riving knife in place on the saw. A splitter or riving knife will help to keep a panel from binding against the saw blade during long rip cuts.

Sheet goods come in a wide variety of compositions to suit a host of different woodworking and construction applications. With proper cutting techniques and sturdy workpiece support, you can cut them safely on the table saw.

Resize large sheets with a portable saw first. If you are using a small jobsite table saw or you don't have a way to support your workpiece adequately around the table saw, start by cutting full-sized sheets down to more manageable proportions with a circular saw and straightedge guide.

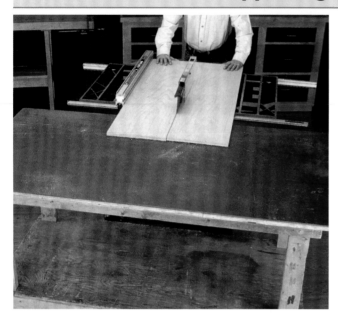

Ripping. When ripping larger sheets, be sure to support the workpiece as the cut is completed. Roller stands will work, but a broad surface, like a shop table, provides even better support.

Crosscutting. Roller stands aren't suitable for supporting long panels when crosscutting. You'll need a support device long enough to span the depth of the saw table and beyond.

Using a Helper for Outfeed Support

Saw operator

Enlist a helper to act as outfeed support if no other temporary support is available. As a workpiece leaves the saw table, a helper should cup the front edge of the board with both hands (but not grip it) and walk backward, keeping pace with the rate you are feeding wood into the saw. It is unsafe for a helper to pull or steer the workpiece; you as the operator must remain in complete control of the cut.

3. Provide Sturdy Workpiece Support

For large panels of sheet stock, workpiece support is crucial. The best setup for cutting full-sized sheets is to outfit a table saw with a large outfeed table butted against or attached to the back of the saw. A 4'–6' or longer outfeed table will provide plenty of support for an 8' length of sheet stock, once it exits the blade. (Be sure the outfeed table is wide enough to accommodate both the workpiece and the waste piece.) You can also use several roller or ball bearing–style stands set behind the saw instead of an outfeed table, but be sure they have stable footings and are lined up squarely with the saw. Roller stands that have cylinders for rollers can skew, tip, or skid along the floor if a heavy piece of sheet stock rolls over them at an angle. A better roller stand for sheet stock uses cupped ball bearings rather than long cylinders for rollers. Ball bearings roll in any direction, so it doesn't matter whether the workpiece contacts them straight on or at an angle.

4. Keep Longer Edges against a Fence

Never rip-cut a panel with the shorter edge against the rip fence, especially if the width is less than 12". The rip fence will not provide enough support for the panel you are cutting, and the workpiece could creep away from the fence, bind against the blade, and kick back. Use a miter gauge outfitted with a long auxiliary fence to crosscut long, narrow panels. Better still, lay them on a shooting board or crosscut sled to make the cut (see top photo, next page).

Stable crosscutting.
Crosscut sleds are a better alternative than a miter gauge and auxiliary miter fence for crosscutting wider panels. Since the sled completely supports the panel from beneath, it cannot pivot or shift as you make the cut. If you are crosscutting panels considerably longer than the saw table, you'll need to provide additional workpiece support alongside the saw, even when using a crosscut sled.

5. Reinforce or Lengthen Your Rip Fence

Larger panels of sheet stock will place more stress on the rip fence as you are making rip cuts. If the rip fence on your saw is T-square style (there is no fence rail along the back of the saw table) or prone to fall out of alignment with the blade, clamp the outfeed end of the fence against the saw table to hold it securely in place.

You can improve the amount of support the rip fence provides by attaching a long auxiliary rip fence (see photo, right). Choose a thick piece of hardwood for the fence to keep it from deflecting. Extending the rip fence is especially helpful if you need to rip large panels of sheet goods on a small jobsite saw. In any case, have the auxiliary fence extend beyond the saw table and onto your outfeed support.

Long auxiliary rip fence

Rip-cutting long panels. Attaching a long auxiliary rip fence will provide more bearing surface when ripping long panels. A long auxiliary rip fence is especially helpful for negotiating sheet goods over a jobsite table saw. For best results, extend the rip fence beyond the back edge of the saw table and onto your outfeed support table.

Use Tape to Minimize Tearout and Chipping on Melamine and Veneer

Sheet goods covered with thin veneers, melamine, or laminate are prone to tearout and chipping around the blade slot opening in the throat plate. To keep this from happening, install a zero-clearance throat plate in the saw (see pages 41–42) and use a sharp triple-chip, crosscut, or plywood blade. Since table saws tend to tear wood fibers on the bottom face of a workpiece where the blade exits the wood, place the "good" side of the panel face on the saw table. If both sides of the panel need smooth, chip-free edges, run a strip of masking tape along the cutting line on the bottom of the panel before you make the cut. The tape will keep the veneer, melamine, or laminate from splintering or chipping.

Table Saw History

All-in-One Shop Tool

Available to carpenters and woodworking shops in the 1920s, this belt-driven, multipurpose machine was outfitted with three table saws—a combination crosscut and ripsaw (right-hand side of engraving) and a swing cutoff saw (left back corner). A truly all-in-one workstation touted by the manufacturer as "eight machines in one," the planing mill was also equipped with a band saw, 12" jointer, tenoner, upright hollow chisel mortiser and borer, reversible spindle shaper, and sanding disk.

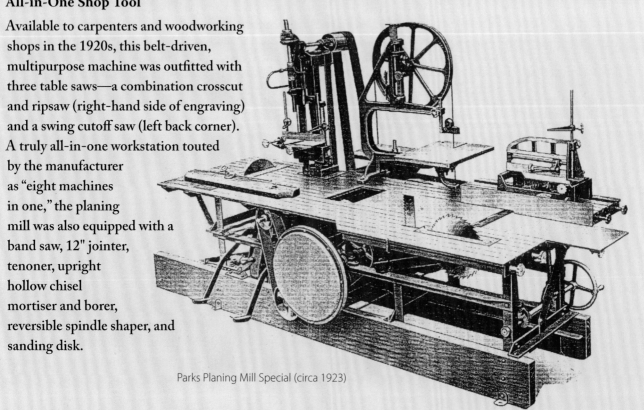

Parks Planing Mill Special (circa 1923)

Ripping Full-Sized Sheets

The procedure for ripping sheet stock is similar to ripping board lumber, but you'll begin the cut farther back and to the left of the saw for cutting full-sized sheets. The object is to feed sheet stock into the blade with your right hand, while keeping the workpiece held against the fence with your left hand.

Start by setting the rip fence as you would for any rip cut. It's a good practice to attach a long auxiliary fence to the rip fence, to increase the rip fence's bearing surface. Extend the fence onto your outfeed table.

Tip

Keep your eye on the joint between the rip fence and the workpiece. It doesn't take much to skew the sheet away from the rip fence, so make any changes in hand or body position gradually. Keep the sheet sliding flat on the saw table at all times.

1 To begin the cut, stand next to the sheet at the left rear corner. Tip the sheet up onto the saw and feed with your right hand. Position yourself behind the sheet near the left corner with your left hand wrapped around the long outside edge, about 1', from the end, and your right hand gripping the end of the sheet at arm's reach. Begin to feed the sheet into the saw, pushing forward with your right hand in a straight line with the blade. Apply diagonal force with your left hand to keep the sheet sliding snugly against the rip fence.

2 Increase your feed rate once the cut has begun, keeping your motions fluid. As the end of the sheet reaches the saw table, shift your body from the left corner around to the back of the panel, moving your right arm so it guides the section of the workpiece between the blade and the rip fence. In those cases where the distance between the rip fence and the blade is wider than about 1', it's safe to use your right hand to feed the workpiece past the blade. Use a push stick in your right hand for narrower workpieces.

3 When the blade cuts the sheet in two, hold the cutoff waste piece stationary with your left hand. Slide the section of workpiece panel between the blade and rip fence completely past the blade with your right hand. Both the waste piece and the workpiece should be supported as they are fed off the saw table with roller stands, an outfeed table, or a helper.

Cutting Cove Molding

You can cut cove molding or curved beveled edges for raised panels on the table saw using a simple adjustable parallelogram jig and a fine-toothed plywood blade. The technique for cutting coves on the table saw involves passing the workpiece over the blade at an angle so the curvature of the blade matches the cove profile you want. In a sense, the table saw blade will act more like a shaping tool than a cutting tool. By varying the angle of approach on the blade, you can machine any number of cove curvatures, from steep recesses with tight radii to gentle, shallow arcs.

Since the saw blade will act as a makeshift shaper and sander when cutting coves, it is important to use a blade with small, rigid teeth. To keep the blade from deflecting due to the lateral forces you'll be applying to the blade, cutting coves will require that you slow down your feed rate when passing workpieces over the blade. In addition, cutting each cove shape involves making multiple passes over the blade, increasing the blade height by not more than ⅛" with each pass, from start to finish.

Build a Parallelogram Jig

You'll need to build an adjustable parallelogram jig from scrap wood to set up cove cuts. The jig takes the guesswork out of determining how a workpiece must pass over the blade to cut a specific cove curvature. Make the long sides of the jig about 4' long and the short ends about 1' long. Attach the parts with short carriage bolts, nuts, and washers. You'll need to be able to swivel the jig so the opening is wider or narrower to accommodate different-sized workpieces.

Draw a curved layout line on the end of your workpiece to mark the cove profile. Raise the blade above the saw table so the teeth at the top of the blade match the highest point of the cove profile.

Adjust the cove parallelogram jig so that the inside edges of the jig match the width of the cove profile. Set the jig on the saw table and over the blade. Turn the whole jig left or right until the inside edges of the jig touch the front and back edges of the blade (see inset photo). In this position, the jig sets the angle you'll need for cutting the cove.

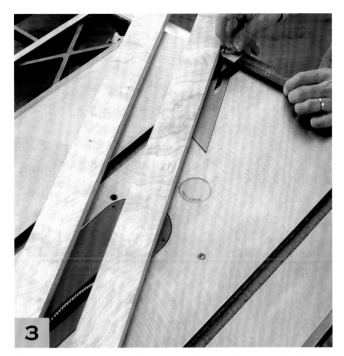

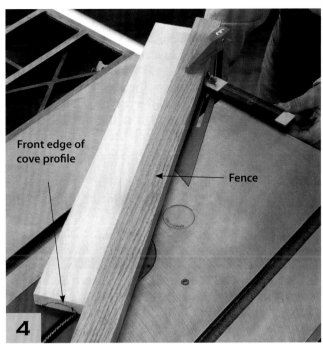

Set a bevel gauge to match the angle the parallelogram jig forms with the front edge of the saw table. Be careful not to bump the parallelogram jig out of alignment as you set the angle on the tool.

Select a length of flat stock about 4' long to serve as a fence for guiding the workpiece when you cut the cove. Set the workpiece and fence on the saw table and against the bevel gauge. Shift the fence until the front blade teeth and the front edge of the cove profile intersect. Clamp the fence in place.

Option: Milling a Partial Cove

Using the workpiece as a spacer, clamp a second fence to the saw table, parallel to the first fence, to create a track for cutting the workpiece. Lower the blade to a height of about ⅛" above the table and make the first cut. Use push sticks to pass the workpiece over the blade. Then cut the cove in a series of shallow passes, increasing the blade height about 1⁄16" with each pass until you've cut away all the material in the cove area.

If your cove profile only needs to be a partial curve shape, cut a saw kerf into the fence you clamp to the saw table so the blade is partially buried in the fence. Cut the partial cove profile as you would a full cove, in shallow 1⁄16" passes.

JOINERY TECHNIQUES

Once your workpieces are cut to size, they'll usually need to be assembled in any of a host of different configurations to form joints. Joints serve three primary functions in a project: They provide flat surfaces on which to apply glue or drive mechanical fasteners; they help bear loads carried by the parts that comprise the joint; and joints allow wood to expand and contract as it is exposed to changes in temperature and humidity.

By design, table saws make excellent, versatile joint-cutting power tools. The saw table, rip fence, and miter gauge fence provide plenty of flat support surfaces for holding a workpiece steady as you cut. Add a few clamps, a simple auxiliary wood fence, or a handful of shop-made jigs, and nearly any flat workpiece can be safely and accurately moved through the blade—on edge, on end, or on its face. You'll need this degree of cutting control for marking joints. in addition, the miter gauge, blade, and rip fence setting can be locked into position, so it's easy to duplicate a joint-cutting procedure exactly and quickly on multiple joint parts.

The purpose of the pages that follow is to present a roster of joints that can be made entirely or almost entirely on the table saw. Many more wood joints exist than are covered here, and some of them simply can't be cut on the table saw. However, unless the hallmark of your woodworking is elaborate joinery, the group of joints covered in this chapter should be sufficient to support and beautify virtually any project you set out to build.

DIFFERENT JOINERY TECHNIQUES

Ways to Join Wood

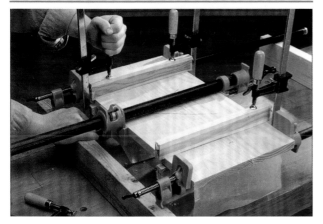

Glue alone is used for some woodworking techniques, including edge-gluing boards together to form wider panels.

Interlocking joints are fashioned with parts that are cut to link together mechanically. Sometimes they are reinforced with nails or screws to reinforce the glue bond.

Pinned or pegged joints, like the mortise-and-tenon joint shown here, feature dowels, splines, or wedges driven across the joint parts to lock them together.

Wood joinery options are almost too numerous to count—which testifies to a long history of innovations in joint craftsmanship. In almost every instance, you'll find that wood is joined in one of three primary ways, all of which can be accomplished with a table saw: Some boards are attached by butting flat faces, ends, or edges together and bonding the joint with glue or mechanical fasteners, such as brads, nails, screws, or hinges. Other joints are made by shaping the mating parts into interlocking forms, then strengthening the union of the parts with glue. A third option for constructing strong wood joints is to insert an additional wooden member (such as a spline, biscuit, wedge, or dowel) between or through the mating parts. Splines and biscuits serve mainly as alignment aids to keep workpieces from shifting while glue joints are drying, but they also add a degree of structural support to the joint. Dowels and wedges typically pin two workpieces together across the joint, which locks them together.

Which joints you choose for a project should depend largely on the look you are after, the strength you need, and the tools you own.

The Basics of Table Saw Joinery

Flat, straight (and usually square) cuts should always be your goal when ripping and crosscutting parts to size. This is especially true when woodworking. A ⅛" gap between parts is less critical if you are cutting plywood roof sheathing, but you'll never build tight furniture joints with this degree of error. The tolerance between these kinds of joint parts is so minimal that parts simply won't fit together unless the mating workpieces are cut precisely. You'll save time and frustration when preparing parts for making joints if you machine your workpieces so they're as flat, smooth, and square as possible.

Caution

Some of the photographs you'll see throughout this book show the blade guard and splitter or riving knife removed. This was done to make the information in the photographs easier to see. To prevent injury, use a working blade guard and a splitter or riving knife with anti-kickback pawls whenever the type of cut being made allows.

Once you've flattened and smoothed those surfaces you'll join together, the next step is to lay out your workpieces. It's important to proportion joints so one half of the joint does not compromise the strength of the other half. For instance, when a groove cut in one workpiece must house another workpiece, it shouldn't be cut so deep as to weaken the rest of the grooved workpiece.

Setting Up Your Saw

It should go without saying that a well-tuned saw greatly improves your prospects for cutting nicely tailored joints. The degree of precision you can expect from your saw is somewhat dependent upon the quality of the saw you own, but even inexpensive portable saws are capable of cutting respectable joints if they are tuned up properly. If you've read the chapter in this book on tune-up and maintenance, you'll recall that there are a few very basic relationships you need to pay attention to when tweaking your table saw. Be sure that your rip fence and blade are parallel and your miter gauge is set square to the blade. Since joinery tends to keep you working in close proximity to the blade (many joints require making repeated short crosscuts), it is important that the throat plate is level with the saw table. Cutting joints can produce small waste pieces that can lodge between the blade and the throat plate opening. In these cases, it's a good idea to install a zero-clearance throat plate (see page 41).

Adjusting for Precise Blade Height

Most cuts made with a table saw are *through cuts* that sever a board in two. Crosscuts, rip cuts, miters, and bevels are examples of these kinds of through cuts. Cutting joints with interlocking parts, however, frequently will require you to make one or more cuts that extend only partway through the workpiece, called *non-through cuts*. Dadoes, rabbets, and slots are examples of this type of cutting operation.

In order to make accurate non-through cuts, you'll need to set blade height precisely to keep the cuts from being too shallow or too deep. A variety of different techniques are used to set blade height. Regardless of the method you choose, always measure off the tooth at the highest point of the blade arc above the saw table. It might be tempting to grab a tape measure to set blade height, but reading off of a flexible tape is difficult because you have to hold the tape in mid-air while you take your reading. A more reliable way to set blade height is to position a try square or combination square on the saw table so it sits perpendicular to the blade, then read the blade height off of the graduations on the square. Be sure the throat plate is flush with the surrounding tabletop, or your blade height reading will be inaccurate. To take a reading, position yourself in front of the saw so your line of sight is even with the height of the blade, and look straight on at the scale next to the blade. Where possible, read off of the scale with the finest increments to get the most accurate reading. If the blade has teeth that bevel to the left and right, rotate the blade, looking for a chipper-style tooth that is ground flat across. It's easiest to get a reliable reading by measuring off this tooth. Otherwise, set the blade height off the tip of a tooth that bevels toward the square.

Options for Setting Blade Height

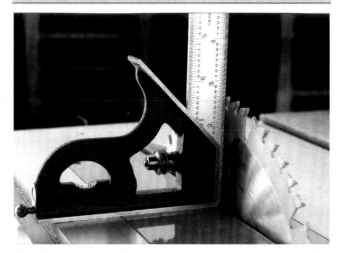

Combination square. Set a combination square on the saw table and against the blade. Use the rule on the square to set blade height. Select the side of the rule with the finest graduations.

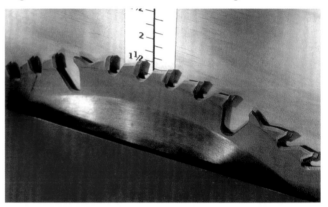

Rip fence scale. Attach a scale to your rip fence at the middle of the blade to serve as a convenient height adjustment reference. Then dial the blade up or down next to the fence to set the height.

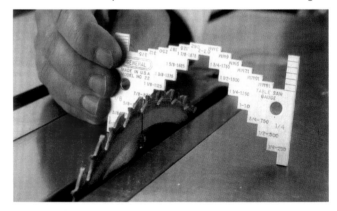

Commercial blade height gauges are stepped off in even increments to take the guesswork out of setting exact blade height. To use the gauge, set it over the blade area and raise the blade until a blade tooth touches the "step" that matches the height you need.

Setting Dado Blade Width

Dado-blade width settings. Stacked dado blades (left): Since the outer blades of a stacked dado blade spin in a single orbit like a standard saw blade, simply measure off the outside edges of each outside blade to determine the edges of the saw kerf. If you'll use the same dado-blade setting to cut a number of different workpieces, you can even mark the "inner" and "outer" edges of the dado blade right on the throat plate. Wobble dadoes (right): To set the width of a wobble dado blade accurately, dial the blade to roughly the width setting you need and make a test cut on a scrap piece. Adjust the blade if needed, and continue to make test cuts until the kerf width on the scrap matches the correct width. Mark the kerf width of the scrap onto the saw's throat plate.

A couple other blade setting options are worth considering. A handy way to set blade height without a square is to mark a vertical scale on your rip fence at the high point where it intersects the middle high point of the blade. Mark the scale with increments as fine as you like, but for practicality and accuracy, ⅛" or ¼" markings are about as close as you can get measuring off the fence. To use the scale, simply slide the fence next to the blade and raise the blade until the highest tooth intersects the mark corresponding to the blade height you need. Various manufacturers also sell plastic or metal blade height gauges for use with table saw blades or router bits (see bottom photo, previous page). The gauges rest on the saw table and are stepped off in even increments. Just raise the blade until it touches the correct "step" on the gauge.

Choosing a Cutting Tool

Every joint covered in this chapter can be cut with a standard-kerf combination saw blade. Regardless of the style of joint, building a joint really is just a matter of making a number of through or non-through rip cuts and crosscuts on edges, ends, or faces until enough material is removed from two workpieces to make them fit together. The good news about cutting joints with a standard blade is that any saw

capable of making reliable straight cuts can cut joints—even the smallest jobsite saw.

A standard-kerf saw blade isn't necessarily the most efficient choice for cutting joinery, however. For instance, if the joint you need to cut is a lap joint where the cutaway area is 2" wide and not on the end of the board, you'd need to make 16 successive crosscuts, removing ⅛" of material at a time to clear away the waste for the joint. Similarly, if the cutaway on the mating piece is also inset on the board, you'll need to make another 16 crosscuts to complete cuts for the joint parts. There's nothing unsafe or incorrect about cutting joints this way, but what a tedious process!

A better way to remove larger amounts of material in fewer passes or even in a single pass is to use a dado blade (see pages 60–62). Most dado blades have a cutting width capacity of nearly 1". That means, at least on shallower cuts and softer woods, you can remove wood at eight times the speed of a standard blade. To be used effectively, however, these tools must be kept sharp and in tune. Keep in mind that the greatest mistake most woodworkers make when using dado blades is to try and remove too much material in a single pass. Once you've become comfortable using a dado blade to accomplish your table saw joinery, you'll never want to go back to using a standard blade for these tasks.

Sawing Joints Safely

Since non-through cuts do not cut a kerf through the workpiece, you'll have to remove the splitter on older saws (riving knives on newer saws can remain in place). The trouble with either a splitter or riving knife, you have to remove the guard. This leaves you more vulnerable to kickback and to direct contact with the blade. The best solution is to replace your saw's standard guard system with an aftermarket guard. On many aftermarket guards on older saws, the splitter is removable, so the guard can remain in place for non-through cuts (see pages 140–141).

If replacing a guard assembly isn't within your budget, you can also construct your own guard systems for protecting yourself against kickback. Sometimes simply clamping a piece of stock next to or on top of the workpiece to hold it in place is all the added security you'll need to perform a cut. As you set up your cuts, do whatever you can to limit the blade's ability to lift the workpiece up and propel it back toward you. When the blade is buried within your workpiece, as it is in a non-through cut, every tooth above the saw table has the potential to come in contact with the workpiece at the same time, which is a prime opportunity for kickback.

Using the Rip Fence and Miter Gauge Together for Non-Through Cuts

Aside from splitter and guard issues, non-through cuts are also unique because in certain circumstances the rip fence and miter gauge can be used in tandem to guide and support workpieces. Remember, crosscuts and rip cuts that sever a workpiece in two parts create waste pieces that can get trapped between the "triangle" formed by the blade, rip fence, and miter gauge. The hazard comes when the skittering, trapped waste piece rubs and catches on the blade teeth on the outfeed side of the blade. In contrast, some non-through cuts do not create waste pieces—only sawdust—so there's no loose projectile produced for the blade to throw back at you. In these situations, the miter gauge serves to support the workpiece as you feed it through the blade, while the rip fence aligns the blade with the layout lines on the workpiece. A number of joint cuts covered in the following pages illustrate how to use the rip fence and miter gauge safely together.

Joint Stresses

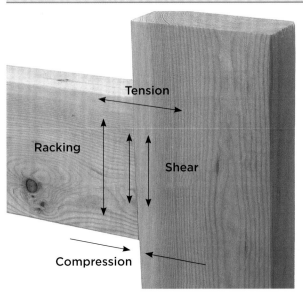

Wood joints undergo four main types of stress: tension, shear, racking, and compression. Each type of joint will vary in its ability to resist failure by one or more of these stresses.

Surface Area

Have you ever wondered why some wood joint styles like dovetails or finger joints have interlocking parts? Other than being visually attractive, the parts that mesh together provide a substantial amount of additional surface area on which to apply glue. On properly made joints, glue surface area is directly related to strength. It isn't the only measure of strength, but it is an important one. For strong joints, the more surface area, the better.

BUTT JOINTS

Butt joints are the simplest woodworking joints to cut and build. When used in the right applications, butt joints are quick to assemble and sufficiently strong. A butt joint is formed when two workpieces are butted together on their faces, ends, edges, or any combination of these surfaces. Unlike other types of joints, the mating surfaces of a butt joint are left flat and smooth, without any additional machining that would allow the pieces to interlock. Instead, the joint typically is reinforced with only glue, so it gains all of its strength from the glue bond.

Because the mating parts do not interlock, butt joints in woodworking situations are inherently weaker than other styles, and they are best left for joining parts that don't need to move or support great amounts of weight. Modern woodworking glues form incredibly strong joints—on a properly made joint, the surrounding wood fibers will break before the glue joint fails, but glue alone won't save a butt joint subjected to excessive twisting, pulling, or shearing forces. If you choose to use butt joints for building drawers or doors, be sure to reinforce the joint with nails, screws, or biscuits where it will be subjected to tension or racking stresses, to keep the joint from pulling apart. Nails, dowels, splines, screws, or bolts will strengthen a butt joint that must carry a load, preventing it from failing due to shear.

Butt joints are used for edge-gluing individual boards together to form wide panels such as tabletops, carcases, workbench, and "butcher-block" surfaces and cutting boards. Edge-glued butt joints are stronger than other butt joint configurations because the boards are joined lengthwise, along the grain (called *long grain*). Long-grain gluing provides plenty of surface area for glue along the joint, and because the wood pores run parallel to the joint instead of across it, the boards can expand and contract evenly across the grain as temperature and humidity fluctuate. Plus, long grain soaks up just enough glue to hold the joint tightly.

A good way to help align long boards in a glue-up is to insert biscuits into the joint. Cut the slots with a biscuit joiner, centering the biscuits across the thickness of the boards.

Another good situation to use butt joints is for face-gluing boards together. This is a common practice for assembling boards into thicker blanks used to make project parts such as table legs and columns. When face-gluing, run the board faces across a jointer or through a thickness planer to ensure that they are flat.

Cutting Butt Joints

Sawing parts for butt joints involves making square, smooth rip cuts and crosscuts. Follow the procedures outlined on pages 63–103 for setting up and making these basic cuts.

A handy technique for sawing smooth, tightly fitting edges is to alternate the desired faces of the boards when running them through the saw (see photos, next page). If your saw is properly tuned and your blade cuts without leaving noticeable saw marks, you can glue butt joints together without needing to plane or joint the edges further. However, jointing the edges can't hurt either long-grain or end-grain glue-ups, so run your workpieces over a jointer where possible, or smooth the mating surfaces with a sharp hand plane. Smoothing the mating surfaces minimizes small gaps that can trap air beneath the glue layer.

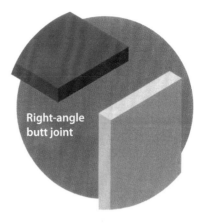

Right-angle butt joint

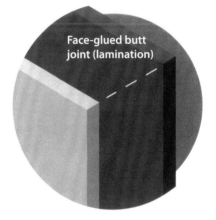

Face-glued butt joint (lamination)

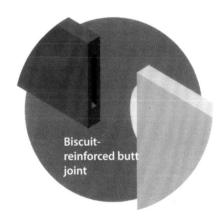

Biscuit-reinforced butt joint

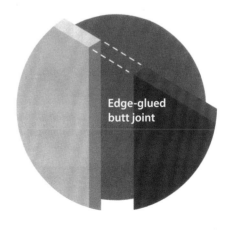

Edge-glued butt joint

Butt Joints

Stress Resistance:

Tension: Poor

Compression: Good

Shear: Poor

Racking: Poor

Common Uses:

Edge-glued panels; light-duty carcase, drawer, and frame construction; rough framing for construction

When you butt the parts together, check the fit of the parts carefully by laying them side-by-side on a workbench. Any noticeable gaps that run lengthwise along the joint indicate that the saw blade wasn't set perfectly square to the saw table for one or both cuts. It's also possible that one of the workpieces may have drifted away from the rip fence slightly, resulting in a wavy cut. You may also have started with a board that isn't straight along its edge. In these situations, you'll need to flatten and square the mating surfaces further before applying glue or clamping up the joint.

How to Cut Mated Edges for an Edge-Glued Butt Joint

1 Lay the two workpieces next to one another in the same orientation in which they'll be fastened. For reference, draw a line or a "V" across the joint.

2 Rip-cut a narrow strip of stock from one of the mating edges. The reference line on the board should be facing up.

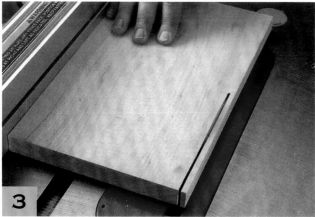

3 Rip-cut a narrow strip of stock from the mating edge of the other board. This time, the reference lines on the board should be facing down.

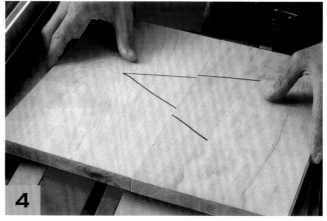

4 Test the fit with the boards oriented in the same fashion as they were when you drew the reference line. Flipping one of the boards should result in a tight perpendicular joint, even if the blade and fence are slightly out of square.

SPLINED BUTT JOINTS

Splines are thin strips of solid wood or plywood inserted into matching slots cut into both members of a joint. On splined butt joints, the slot runs along the joint, centered on the thickness of the mating workpieces. Adding a spline between the mating surfaces of a simple butt joint is an easy modification to make and offers several advantages over butt joints bonded with glue alone. First, splines act like dowels or biscuits to help interlock and align members of a joint during glue-up.

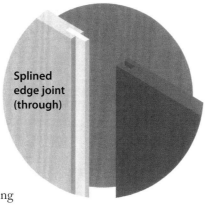

Splined edge joint (through)

Splined edge joint (concealed)

Cutting spline slots along the length of a butt joint also contributes to a stronger glue joint, since the spline slot more than doubles the surface area for glue; increasing the gluing area helps strengthen the joint against forces of tension, stress, and racking. Plus, the spline acts almost like a tenon, making the joint more resistant to failure caused by shear forces. A spline joint, however, isn't as strong as other two-part interlocking joints that will be covered in this chapter.

Right-angle splined joint

Cutting Spline Joints

The width of a spline is typically equal to the thickness of the mating parts, so you'd use a ¾"-wide spline for 1x stock. Spline thickness should not exceed one-third the thickness of the mating pieces. For gluing up ¾"-thick stock, splines ranging from ⅛" to ¼" thick are suitable. Since splines serve primarily as alignment aids, splines wider than ¾" do not contribute more strength to the joint. When gluing up boards thicker than ¾", a better alternative is to install two splines side by side in the joint.

To cut a spline slot, install either a standard-kerf saw blade or dado blade in the saw and set the blade height to one-half the spline width plus about 1⁄32" (the extra height allows for excess glue). If using a dado blade, you'll also need to set the cutting width. If you don't own a dado blade but need to cut a spline slot wider than the width of your saw blade, make multiple passes with a standard blade.

Lay out slot locations on each member of the joint. Splined butt joints are strongest when the splines are centered on the thickness of the mating parts. Then set the rip fence so the saw blade aligns with the layout marks on your workpieces. Cut the slots for the parts. Use a tall fence to support wide workpieces when cutting on edge, and use a tenoning jig for cutting the ends of long, narrow boards. Install featherboards to hold your workpieces tight against the rip fence when possible.

To make sure the slots on mating boards align, mark the board faces and run the marked faces through the saw in the same orientation.

Cutting Splines

To make splines, measure the width of the spline slot carefully to set up the cut. Rip-cut splines from the edge of a board so the splines fall away from the blade on the side opposite the rip fence. If you want to hide the spline ends, cut concealed slots or make the splines from the same wood species as the workpieces. For a decorative highlight, choose wood with contrasting color.

To cut several narrow splines quickly and easily, use a narrow ripping jig (see page 76).

Splined Butt Joints

Stress Resistance:

Tension: Fair

Compression: Good

Shear: Fair

Racking: Fair

Common Uses:

Edge-glued panels; "free-floating" cabinet panels; carcase, drawer, and frame construction

How to Edge-Glue with Splines

1 Set the blade height to one-half the width of the spline, plus ¹⁄₃₂". Make a lengthwise kerf cut on each mating edge.

2 Measure your saw blade kerf, then rip splines to that thickness. Test the fit. If necessary, adjust the saw setup and cut new splines— don't try to trim splines that are too thick.

3 Apply wood glue in one of the slots, then insert the spline. Apply glue in the mating slot and slip the mating board over the spline. Draw the workpieces together with clamps. Do not over-tighten the clamps.

How to Cut a Concealed Spline Joint

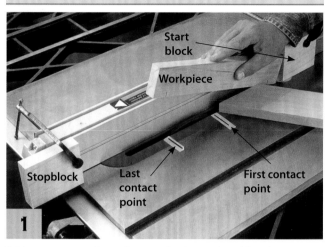

Start block

Workpiece

Stopblock

Last contact point

First contact point

1 With your saw unplugged, set up for the cut by raising the saw blade to the cutting depth (usually ³⁄₈"), then marking the points of first and last contact with blade onto the saw table. Make a reference mark 1" from each end of the workpiece, then clamp a wood block to the saw fence, positioned so the reference marks will align with the appropriate blade contact marks at the start and finish of the cut.

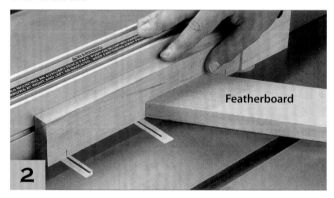

Featherboard

2 Using a featherboard to hold the workpiece against the fence, turn on the saw and carefully lower the workpiece onto the spinning blade. The back end of the workpiece should be butted flush against the start block.

3 Feed the workpiece through the blade, guiding it with a push stick. When the front end contacts the stopblock, switch off the saw and remove the workpiece once the blade stops spinning.

RABBET JOINTS

Rabbets (sometimes called *rebates*) are rectangular cuts made along the edge or end of a workpiece to form a tongue. Rabbet joints are made in two styles. In the case of *double rabbet* and *shiplap* joints, a rabbet is cut into the mating ends or edges of both workpieces, and the proportions of the rabbets match one another. This way, the rabbeted parts can be combined so the workpieces form a right angle (double rabbet) or a flat plane (shiplap). The second rabbet joint style, called an *overlap rabbet*, involves cutting a deep rabbet in one workpiece so it overlaps the square end or edge of a mating workpiece.

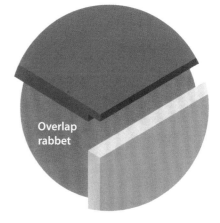

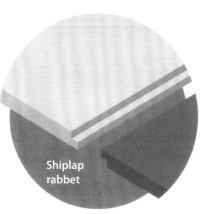

Rabbet joints are a better alternative to butt joints for building boxes and carcases because the joints partially interlock. You'll commonly see double rabbets and overlap rabbets used in bookcase construction to conceal the ends and edges of the back and top panels. Depending upon how the joints are arranged, both of these rabbet joint types can conceal most of the end grain. Double and overlapping rabbet joints typically are bonded with glue and reinforced with brads or nails.

Shiplap joints are superior to edge-glued butt joints when gluing up panels because the "stepped" pattern of the joint lends one-third more surface area for gluing, creating a stronger joint. When the rabbets are cut accurately, a shiplap self-aligns the workpiece faces and resists shifting during glue-up.

Cutting with a Standard Blade

Cutting rabbets with a standard-kerf saw blade involves cutting the cheeks with one saw setup, then trimming the shoulders in a second operation. For double and shiplap rabbet joints, mark cheek and shoulder layout lines on both workpieces, checking the rabbet layouts against one another to be sure the proportions will match. It doesn't matter which cut you make first—cheeks or shoulders—but it is most efficient to make like cuts on all workpieces before you change fence or blade settings to make the second set of cuts. Also, be sure that the waste piece is on the side of the blade opposite the fence so it does not become trapped between the fence and the blade when it falls free.

To make cheek cuts, you'll need to stand workpieces on edge or end against the rip fence. Install a tall fence for rabbeting the edges of wide workpieces, and use a tenoning jig to support workpieces with cheek cuts on the board ends. When setting the distance from the rip fence to the blade, be careful that the blade will cut inside the waste area of the rabbet, rather than into the tongue of the rabbet (see top photos, next page). Set the blade height so the teeth just touch your shoulder layout marks. Make a test cut on scrap to check your fence and blade settings.

For rabbets that follow the edge of a workpiece, cut the shoulders using the rip fence as a guide. Rabbet shoulders cut into the end of a board should be cut with the miter gauge as a guide. Be sure to reset the blade height for cutting shoulders on overlap rabbet joints, because the shoulders are shorter than the cheeks.

Double-Rabbet Joints

Stress Resistance:

Tension: Fair

Compression: Good

Shear: Fair

Racking: Fair

Common Uses:

Light-duty drawer construction; carcases and boxes; edge-glued panels

How to Cut a Rabbet with a Standard Blade

1

Draw layout lines for the rabbet on the workpiece. Set the blade height and ripping width to the dimensions of the rabbet, then make the cheek cut. Use a featherboard to hold the workpiece against the fence. To prevent the waste piece from becoming trapped in the throat opening, use a zero-clearance throat plate (see page 41).

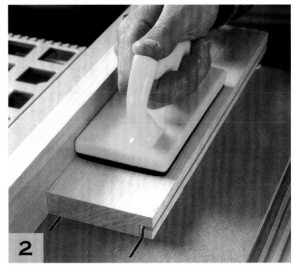

2

With the rip fence positioned so the waste piece will be on the outside of the blade, make the shoulder cut. Use a push pad to guide the workpiece over the blade.

Cutting with a Dado Blade

Dado blades make quick work of cutting rabbets, because the blade cuts the cheek and shoulder in one pass. Regardless of the dimensions of the rabbet, orient your workpieces to lie facedown on the saw table when you make rabbet cuts with a dado blade. The width of the dado will set the height of the rabbet cheek, and the height of the blade will determine the depth of the shoulder. If the depth of the shoulders exceeds the maximum height of the dado blade, you'll have to make the cheek and shoulder cuts with another saw, such as a band saw or a reciprocating saw.

Install a sacrificial wood fence on the rip fence and set the dado blade to cut wider than needed to cut the rabbet. Raise the dado into the sacrificial fence so part of the blade will be shrouded by the sacrificial fence. Set the blade height and fence distance from the blade to match the cheek and shoulders of your rabbet. For rabbets that follow the edge of a workpiece, make the cut by guiding the workpiece along the rip fence and over the dado.

Dado blades create more resistance against a workpiece than a standard saw blade, and the resistance increases the more material you remove in one pass. Clamp a scrap hold-down block or featherboard above the workpiece to help hold the workpiece against the saw table as you cut with a dado blade.

When cutting a rabbet along the end of a workpiece that's narrower than about 8", use the miter gauge to support the workpiece from behind as you feed it over the blade. It is safer to use the miter gauge in tandem with the rip fence in this instance, rather than simply pushing the workpiece

through by hand, because the miter gauge provides broader support and keeps the workpiece from shifting. Without the miter gauge, you would be forced to run the narrow edge of the workpiece tight against the fence.

How to Cut a Rabbet with a Dado Blade Set

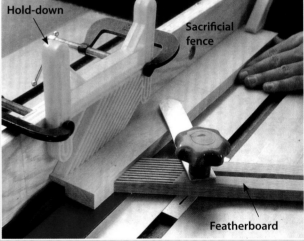

Hold-down

Sacrificial fence

Featherboard

A dado blade can cut shallow rabbets in a single pass. If the rabbet is deeper than ⅜" or the workpiece is hardwood, make the cut in multiple passes. Attach a sacrificial wood fence to the rip fence. Set the fence position and cutting depth, then make the cut. Use a hold-down, featherboard, and push stick to guide the workpiece.

RABBET-AND-DADO JOINTS

Rabbet-and-dado joints combine the enhanced gluing area of a rabbet with the interlocking characteristics of a dado or groove to form joints that are very resistant to failure from racking or shear. Rabbet-and-dado joints take two forms, depending on their intended purpose. One style features a rabbet with a tongue half the thickness of the workpiece that fits into a matching dado or groove. This joint is used commonly to fasten shelves to upright standards, such as bookcases. The ends of the shelves receive the rabbets, while the standards are dadoed. Another less common use of this joint is for frame-and-panel doors—the rabbet is cut around the four edges of the panel and the dado (actually a groove in this case) is centered around the inside edges of the frame. The panel side with the rabbet tongues faces out, which leaves a decorative "reveal" between the front of the frame and the panel. The advantage to this frame-and-panel style is that the panel can set back from the front face of the frame without needing to be beveled or planed in order to fit in the narrow dado.

A second form of rabbet-and-dado joint, often called the "bare-faced" tongue and dado, features a thin rabbet tongue that fits into a ¼"-deep dado in the face of the mating workpiece. The tongue can be as thin as ⅛" but should not exceed one-fourth the thickness of the workpiece. This joint is primarily used to form corner joints in drawers. Rabbets are cut along the ends of the drawer front or back, and dadoes are cut in the drawer sides. Since drawer joints are located either in or close to the ends of the drawer parts, the thin tongue side of the joint faces into the drawer. Positioning the tongue farther into the drawer sides keeps the tongue from weakening the short-grain material between the tongue and the ends of the drawer sides.

Cutting Rabbet-and-Dado Joints

Shelf-style rabbet-and-dado joints with thick rabbet tongues can be cut with a standard-kerf saw blade. The rabbet half of the joint is cut using the same two-step process as was described for cutting rabbet joints with a standard saw blade (see page 114). However, cutting the dado will require several passes with a standard blade to reach the final dado width. For this reason, it's more efficient to cut these joints with a dado blade. Cut the rabbets first, then size the dadoes to match. Dadoes running across the grain should be cut using the miter gauge as a guide. If the dado runs lengthwise and with the grain (called a *groove*), cut it by running the workpiece along the rip fence.

"Bare-faced" tongue-and-dado joints also can be cut with a dado blade when the thickness of the tongue is equal to or exceeds the minimum width setting of the dado blade (usually ³⁄₁₆"). Use a standard saw blade with a ⅛" kerf for cutting joints with thinner tongues. Here's how: Mark the dado-side workpieces and set the blade height to ¼". Cut the dadoes by feeding the workpieces over the blade and against the rip fence. Support the workpieces from behind with the miter gauge if sawing across the grain. Next, cut the rabbet tongues in the other workpieces, standing them on end against a tenoning jig. Since you'll be cutting with the blade only ⅛" from the fence, attach an auxiliary fence to the rip fence and install a zero-clearance throat plate. Lower the blade about ¹⁄₃₂" from the setting used to cut the dadoes. Once you've made the cheek cuts for the rabbet tongue, lay the workpiece flat on the saw table to cut the rabbet shoulder.

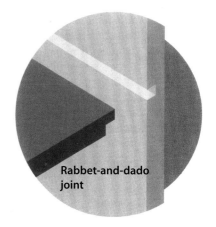

Rabbet-and-dado joint

Bare-faced tongue and dado

Rabbet-and-Dado Joints

Stress Resistance:

Tension: Fair

Compression: Good

Shear: Good

Racking: Good

Common Uses:

Drawer and bookshelf construction; carcases; boxes

1

Lay out and cut the dado. We used a dado blade to cut the ¼"- wide dado for the drawer back shown above. We used the rip fence as a stopblock to index the cut and guided the workpiece with the miter gauge (because no through cut is being made, there is no risk of the waste piece binding).

2

Make the cheek cut for the rabbet. We used a shop-made tenoning jig (see pages 162–163) to secure the workpiece as we made the cut.

3

Finish the rabbet by making the shoulder cut to remove the waste from the rabbet area. A miter gauge with an auxiliary fence is being used to guide the cut.

4

Assemble the joint. For most purposes, glue alone will provide adequate reinforcement.

TONGUE-AND-GROOVE JOINTS

Tongue-and-groove joints combine a number of important features to create a strong, easy-to-build alternative to butt or rabbet-and-dado joints. Tongue-and-groove joints run along the grain. Both joint types are fashioned by cutting a centered tongue in the edge or end of one board that fits into a centered dado or groove on the mating workpiece. As with splined butt joints, the tongue helps align parts during glue-up. By centering the tongue and fitting it into a centered dado or groove, the shoulders on the tongue side of the joint completely hide the slot made by the dado or groove—a real advantage for concealing the slot side of the joint. The tongue is cut to one-third the thickness of the workpieces, so it strengthens the joint against racking stresses. Centering the tongue along the joint also adds to the overall surface area of the joint.

These centered tongue joints find their way into a variety of different panel and carcase applications. Tongue-and-groove joints are often used to form cabinet back panels, and the pieces are assembled without glue. This way, the panel can expand and contract freely, and the tongues hide gaps that would otherwise show if the parts were merely butted together. Tongue-and-groove or tongue-and-dado joints also are used to build frame-and-panel doors and in drawer carcase construction, but neither of these applications produce the strongest door or drawer joints possible. A better alternative for drawers is the finger joint (see page 130) or the double-dado joint (see page 122). Mortise-and-tenon joints form the strongest frames (see page 126).

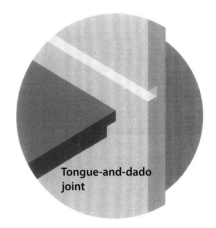

Tongue-and-dado joint

Tongue-and-groove joint

Note

If the groove is cut across the width of a board, the joint is properly known as a *tongue-and-dado*.

Cutting Centered Tongue Joints

Cutting tongue-and-groove or tongue-and-dado joints is a three-step process. You'll first cut the groove or dado along the edge, end, or face of one workpiece, then make two identical rabbets along the edge or end of the mating workpiece to form the tongue. It doesn't matter whether you cut the slot first and then the tongue or the other way around, but accuracy is important either way to ensure that the parts fit together snugly without binding. These joints can be cut most efficiently with a dado blade, but you could also use a standard saw blade following the multiple-step process for cutting rabbets and dadoes described on page 114.

To cut the slot, mark the dimensions of the dado or groove onto the face, end, or edge of the workpiece. Slot width should equal one-third the thickness of the tongued workpiece. If the slot will be cut into the face of a board, make its depth equal to half the board's thickness. Tongues that run along ends or edges are typically ½" long, so set end and edge slots to this depth. Attach a sacrificial wood fence to your rip fence (you'll need it for cutting the tongue) and set the width and height of the dado according to your layout dimensions. Cut the slot in one pass. Use a tenoning jig to support workpieces when cutting grooves that run along the narrow end of a workpiece. For grooves cut along one of the long edges of a workpiece, be sure the sacrificial wood fence is tall enough to provide adequate support as you slide the workpiece along the fence and over the blade. Hold the work snugly against the fence with a featherboard. Use the miter gauge in conjunction with the rip fence when cutting dadoes across the face of a workpiece.

You'll want to cut the tongue slightly shorter than the slot depth so the shoulders of the joint will seat tightly against the slotted workpiece. To account for a shorter tongue, lower the dado blade ⅟₃₂" to ⅟₁₆" from the setting you used to cut the slot. Reset the dado width so it's wider than the final tongue width (a

Tongue-and-Groove/ Dado Joints

Stress Resistance:

Tension: Good

Compression: Good

Shear: Good

Racking: Good

Common Uses:

Shelving; solid-wood flooring and paneling; edge-glued or "floating" panels; "bread-board"-style edges and ends for tabletops; drawer carcases; frame-and-panel doors dado-blade width of ½" is sufficient for cutting the rabbets for tongues in 1x stock). Set the sacrificial rip fence over the dado blade so the exposed area of the blade matches the shoulder width of the tongue (¼" for tongues in 1x stock).

Mark the tongue layout on both ends or edges of a scrap piece that matches the thickness of the actual workpiece. This way you can test the fit of the tongue in the slot you've already cut and make adjustments to the dado setup before committing to your final workpiece. To machine the tongue, cut one rabbet along the edge or end of the scrap, flip the board end-for-end (or edge-for-edge), and cut the second rabbet. Support the rabbet cuts appropriately for the specific joint you are building, using tenoning jigs, tall rip fences, or the miter gauge. If the resulting tongue is too long and holds the joint apart, correct the problem by lowering the dado blade slightly and retest on fresh scrap. Reset the fence forward to cut a thicker tongue or back to trim the tongue thinner. Remember, even a slight adjustment will magnify the changes to a tongue, since you are essentially adding or subtracting thickness by duplicating the same cut on each side of the tongue. Make additional test cuts on scrap until you achieve a good tongue-and-slot fit, then cut your workpieces.

How to Make a Tongue-and-Groove Joint

1

Cut a groove in the edge of one mating board, using a dado blade or by making multiple passes over a single blade. The groove should be one-third the thickness of the workpiece.

Check tongue setting by cutting test board before cutting actual workpiece

2

On the uncut mating board, remove the waste wood on one side of the tongue area. Any rabbet-cutting technique may be used with either a standard blade or a dado blade.

3

Flip the workpiece and remove the waste on the other side of the tongue.

4

Apply wood glue in the groove, then insert the tongue. The tongue should fit fully in the groove.

HOUSED DADO JOINTS

Housed dadoes are a quick alternative to cutting either rabbet-and-dado or centered tongue joints when attaching shelves to uprights. To construct the joint, you simply cut a dado that matches the full thickness of the mating board (typically, the shelf). Shelves built with housed dadoes have good shear resistance, but the longer the shelf, the more likely it will twist out of the dado if it should bow. In situations where appearance is not important, reinforce the glue joint by screwing or nailing through the dado and into the housed board. This will also make the joint more resistant to failure from tension stresses. If appearance matters, conceal the finished joint with iron-on veneer tape or solid-wood edging.

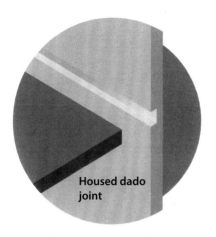

Housed dado joint

Cutting a Housed Dado

The best blade for cutting housed dadoes is a dado blade. When cutting the dadoes, back up the cuts with an auxiliary fence attached to the miter gauge. Otherwise, the dado blade could tear out material on the back side of the workpiece. Lay out the dadoes so the dado depth does not exceed one-half the thickness of the workpiece. Make test cuts in scrap first to check the fit of the parts.

A good practice when cutting housed dadoes for shelving is to cut the dadoes across both shelf standards simultaneously. Do this by cutting dadoes across one workpiece, then ripping the workpiece in half lengthwise to form two identical shelf standards. This way, when you assemble the shelving unit, the shelves will line up perfectly and parallel with one another between the standards.

Housed Dado Joints

Stress Resistance:

Tension: Fair

Compression: Good

Shear: Good

Racking: Good

Common Uses:

Attaching shelving to carcases and drawer backs to sides

Tips for Cutting Housed Dadoes

Metal or paper shims

Find the right width. If you are using a stacked dado blade to cut a housed dado, be sure the dado width is precise so the housed piece will fit tightly into the dado. Experiment with different groupings of chippers and shims until the exact cutting width is attained.

Guaranteed shelf alignment. When cutting housed dadoes in bookshelf standards, cut the dadoes across a wide workpiece, then rip the workpiece in half to create two identical standards.

BLIND DADO JOINTS

Blind dado joints have the same structural characteristics as housed dadoes, but the dado stops short of running completely through to the front of the workpiece. The joint is concealed on the finished side without the need for veneer tape or wood edging. A corner of the housed member of the joint is notched so it fits into the end of the dado, but it still allows the part to extend all the way to the finished side of the assembly.

Cutting a Blind Dado

Stopping a dado cut on a table saw at a specific point involves establishing a reference point on the saw table or fence to mark where the cut should end on the workpiece. You could set this reference by marking on a piece of tape, but this method requires you to stop the cut by visually aligning the edge of the workpiece with the marker. A more accurate and foolproof method is to clamp a stopblock to the saw table. This way, the reference marker is also a physical barrier to keep you from accidentally cutting too far.

To set up the stop-block for cutting blind dadoes, set your workpiece against the miter fence and slide the workpiece forward until the leading edge of the board touches the dado-blade teeth on the infeed side of the blade. Determine the length of the dado cut by measuring across the width of your workpiece and subtracting the amount by which you wish to stop the dado short of the workpiece edge. Without moving the workpiece, measure this distance, starting from where the workpiece touches the blade. Clamp a stopblock in place at this point. Be sure to clamp the stopblock firmly to the saw table to keep it from shifting during the cut, especially if you are making multiple matching cuts on your workpieces.

If you've calculated correctly, the dado cut should terminate exactly where you want when you make an initial cut on a scrap piece. To cut closer to the edge of the board, adjust the stopblock so it's closer to the blade. To stop the cut farther from the edge of the workpiece, adjust the stopblock so it's farther from the blade. Once you've cut the dadoes, square up the ends of the stopped cuts with a chisel.

Scribe layout lines for the notches on the corners of the workpieces (usually shelves) that will be housed in the dadoes. The dimensions of the notches should match the depth of the dadoes and the distance by which the dadoes stop short of the front edge of the workpiece. Cut the notches by standing the shelves on edge against the miter gauge and a tall auxiliary miter fence. Inset the dado blade in a sacrificial rip fence so the width of the exposed blade and its height above the saw table match the layout lines of the "shelf" notches. Set the end of each shelf against the rip fence and use the miter gauge to support the workpiece as you feed it through the blade.

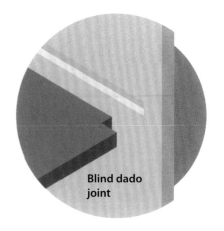

Blind dado joint

Blind Dado Joints

Stress Resistance:

Tension: Fair

Compression: Good

Shear: Good

Racking: Good

Common Uses:

Shelf support

How to Cut a Blind Dado Joint

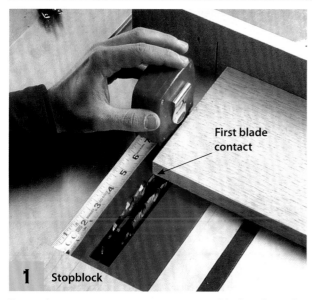

1 Stopblock

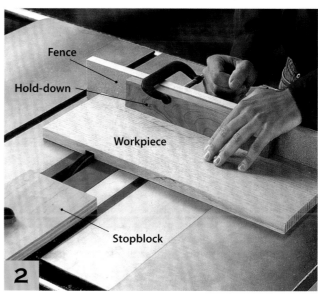

Fence

Hold-down

Workpiece

2 Stopblock

Use a rule or tape measure to position a stopblock at the end of the cut, allowing for the length of the workpiece.

With a fence attached to your miter gauge, clamp a hold-down over the back of the workpiece. Make the cut.

3

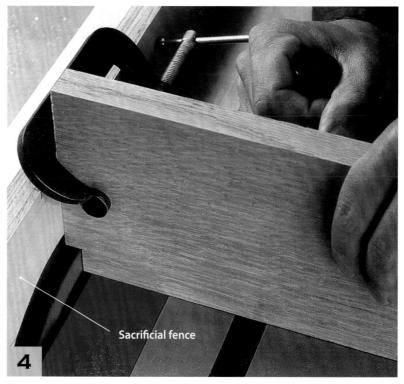

4 Sacrificial fence

Clean out and square off the stopped end of the blind dado with a sharp chisel (preferably, the same width as the dado).

Notch the un-dadoed workpiece (typically, a shelf) at both ends to fit around the front edges of the dadoed workpieces. A dado blade that's partially recessed into a sacrificial fence is perfect for cutting notches.

DOUBLE-DADO JOINTS

Double-dado joints are a hybrid cousin of the tongue-and-groove joint, but they can be a little befuddling to the eye when you study how the parts fit together. Unlike standard tongue-and-groove joints, double dadoes actually have a tongue on each member of the joint, rather than a tongue on one part that fits into a dado or groove on the other part. Notice that the tongues on both joint parts are off center and the same length. When you fit the joint together, the tongues fit into dadoes in the opposite workpiece.

By virtue of their interlocking configuration, double dadoes are resistant to nearly all joint stresses and provide a large surface area in and around the tongues and grooves for glue. They are also a good choice for applications where exposed end grain is undesirable. These joints could be used for carcase construction, but you'll find them most commonly reinforcing drawer frames. The side of the joint that conceals the end grain of the other joint part becomes the front of the drawer. This way, both the interlocking tongues and the glue bond strengthen the joint against stresses associated with opening and closing the drawer.

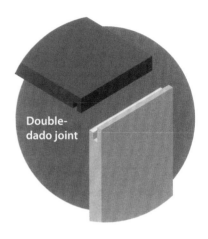

Double-dado joint

Cutting Double-Dado Joints

The best procedure for cutting double dadoes is to cut the "drawer side" dado first to establish the thickness and length of one tongue, then cut the dado on the end of the "drawer front" and trim the tongue in one or more passes until the tongues and dadoes interlock and the drawer side fits fully into the drawer front.

If two drawer parts are each ¾" thick, set your dado blade to ¼" wide and ⅜" cutting depth. Set the rip fence ¼" from the blade. Cut the drawer side dado on scrap stock first to check your setup. Since you are cutting the dado on a narrow end of the board, back up the cut with an auxiliary fence attached to the miter gauge. Adjust the setup as necessary and cut the drawer side dado in the actual workpiece. If you plan to use the same joint all around a drawer, cut the other three dadoes in the two drawer sides at this time.

Cut the dadoes in both ends of the drawer front by clamping the drawer front on end in a tenoning jig. Set the depth of the dado to match the thickness of the drawer side. Trim the inside tongues on the drawer front to length until the drawer side fits completely into the drawer front (see step 3, next page). To trim the tongues, lower the dado blade so it just trims off the tongue without cutting into the other tongue.

Double-Dado Joints

Stress Resistance:

Tension: Fair

Compression: Good

Shear: Good

Racking: Good

Common Uses:

Drawer construction

Note

Double dadoes are usually cut on the drawer front only. The drawer back, which is subjected to less stress, is held in place with housed dadoes.

How to Cut a Double-Dado Joint

1

Cut the first dado in one of the "drawer side" workpieces using a dado blade or by making multiple passes with a standard saw blade.

2

Use a tenoning jig to stabilize the "drawer front" workpiece as you cut a dado in the ends of the board.

3

Trim off the top, inside face of the "drawer front" workpiece to create a short tongue that will fit into each dado in the drawer sides.

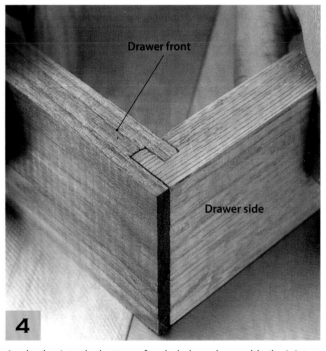

Drawer front

Drawer side

4

Apply glue into the bottom of each dado and assemble the joint.

LAP JOINTS

Lap joints combine the interlocking characteristics of rabbets and dadoes in a number of different configurations to join two boards face to face. Lap joints are a better alternative to butt joints for frame construction, because laps offer much more gluing area and the pieces actually interlock mechanically. Usually, the thickness of both parts of a lap joint are equal, and the pieces fit together so that both faces of each joint part line up with one another. In order to do this, the overlapping dadoes or rabbet tongues are as large as the mating piece is wide.

Several lap joint styles can be cut entirely on the table saw: corner half-laps, "T" half-laps, cross half-laps, full laps, angled half-laps, and bridle joints. "Half-lap" joints get their name because the dadoes or rabbets cut for the joint extend half the thickness of each workpiece. Since the interlocking portions of half-lap joints match, the same blade setup can be used for cuts on both workpieces, which reduces the number of setups needed to cut the joint.

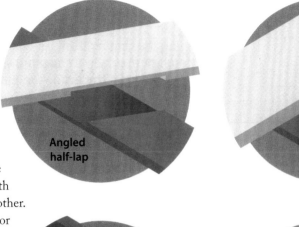

Angled half-lap

Bridle joint

Corner half-lap

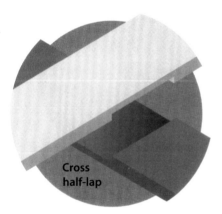

"T" half-lap

Cross half-lap

The procedures for cutting lap joints on a table saw are similar, regardless of the specific joint. Corner half-laps consist of two wide rabbets cut across the faces of two workpieces at the board ends. The cheek portion of each rabbet matches the width of the opposite workpiece. "T" half-laps fasten the end of one board with a wide rabbet to the midsection of another board that's fitted with a matching dado. Full lap joints resemble "T" half-laps, but the dado is actually deep enough to house the full thickness of the board joined on end so it needs no rabbet. "T" half-laps, cross half-laps, and angled half-laps all are made by cutting two wide dadoes into the joint members—either on the faces or edges—then interlocking the dadoes.

Cutting Lap Joints

Aside from their larger size, the rabbets and dadoes of lap joints are cut just like any other rabbet or dado. Whatever blade you choose, you'll need to make multiple passes to cut these rabbets and dadoes. For the most efficient cutting, it makes sense to use a dado blade so you can remove more material with each pass. A dado blade is also the right choice for long or unwieldy workpieces. In these instances, cut the rabbets or dadoes in multiple passes with the workpiece lying facedown on the saw table. Be sure to guide only manageably sized workpieces in a tenoning jig whenever you need to cut rabbets with the workpiece standing on end.

Double-Dado Joints

Stress Resistance:

Tension: Good

Compression: Good

Shear: Good

Racking: Fair

Common Uses:

Frame and face frame construction; furniture leg stretchers

Tips for Cutting Lap Joints

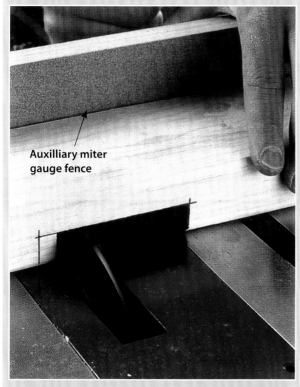

Auxilliary miter gauge fence

Use a dado blade to remove waste material quickly from notches used in lap joinery.

Cut angled half-laps by adjusting the protractor head on the miter gauge. Note the sandpaper adhered to the auxiliary miter gauge fence for extra grip on the board.

How to Cut the Tongue for a Half-Lap Joint

1

We used a crosscut sled (see page 91) to guide the workpiece when making the initial shoulder cuts for the tongue portion of this half-lap member. Cut halfway through the thickness of the board at a cutting width that matches the width of the mating workpiece (below).

2

Use a tenoning jig to stabilize your workpiece when trimming away the waste in the cheek area of the tongue.

Tip

Use the workpiece as a reference for setting up the width of the cut of the mating tongue.

MORTISE-AND-TENON JOINTS

Mortise-and-tenon joints are traditional joints for fine furniture making. You won't find mortise-and-tenon joints in many mass-produced pieces of furniture or cabinets these days, especially with modern quick fastening options like wood biscuits and pocket screws increasing in popularity. But whenever strength and stress resistance are critical and quality is a concern, you'll likely find a mortise-and-tenon joint.

Mortise-and-tenon joints take full advantage of those factors that contribute most to joint strength: namely, large gluing surface area and parts that interlock. Have a look at the various types of mortise-and-tenon joints illustrated on this page. You'll see that tenons have from two to four cheeks and shoulders, and all of these cheek and shoulder faces contribute surfaces for glue. As a rule of thumb, tenons are usually one-third the thickness of the workpiece and up to 3" long. These proportions maximize the shear strength of the tenon without compromising the strength of the mortise walls on the mating workpiece. A properly constructed mortise-and-tenon interlocks snugly without binding, just loose enough to permit a thin, even glue bond along the full joint.

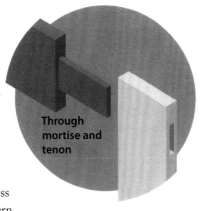

Through mortise and tenon

Open mortise and tenon

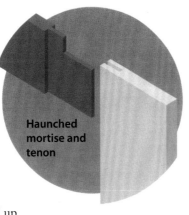

Haunched mortise and tenon

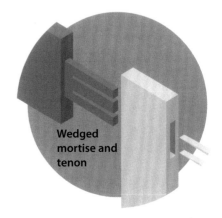

Wedged mortise and tenon

Cutting Mortise-and-Tenon Joints

Only one style of mortise-and-tenon joint, the open style, can be cut entirely on the table saw. The mortise is situated on the end of the workpiece, and one end of the mortise is open, so the mortise slot can be cut on the table saw if the workpiece is fed through the blade with the help of a tenoning jig. The tenon tongue fits into the open mortise on the edge or end of the mortise workpiece, so the joint resembles an elongated tongue-and-groove. Other mortise-and-tenon joint types all incorporate a mortise that is hollowed out from the workpiece. These mortises must be excavated with a mortising chisel, drill bit in a drill press or right-angle drilling jig, plunge router, or dedicated mortising machine with special hollow chiseling bits.

The process for cutting tenons on a table saw is straightforward, simple, and possible with either a standard saw blade or a dado blade. Which blade you choose depends mainly on the number of tenons you need to cut and the proportions of the workpiece.

Cheek and shoulder cuts can be made with a standard saw blade in two ways: First, you can cut the tenon with the workpiece face lying on the saw table by making multiple, side-by-side shallow passes to clear away one cheek at a time. Clamp a stopblock to the miter gauge to start the cheek cuts at the shoulders, and form each cheek by working toward the end of the workpiece. Then flip the workpiece to the opposite face, cut the second shoulder, and proceed with cutting the second cheek. If the tenon has four cheeks and shoulders, stand the workpiece

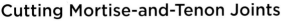

Mortise-and-Tenon Joints

Stress Resistance:

Tension: Good

Compression: Good

Shear: Good

Racking: Good

Common Uses:

Frame, face frame, and door construction; joining table aprons and chair rails to legs

Caution

Some of the photographs you'll see throughout this book show the blade guard and splitter or riving knife removed. This was done to make the information in the photographs easier to see. To prevent injury, use a working blade guard and a splitter or riving knife with anti-kickback pawls whenever the type of cut being made allows.

on edge against the miter gauge to cut the narrow cheeks. You may need to change the blade height for cutting the narrow cheeks, but the stop-block position doesn't change.

Another method for cutting tenons with a standard blade is to cut each cheek in one pass with the workpiece standing on end against the tenoning jig. Set the blade height so the tips of the teeth cut to the shoulder line, and be sure to account for the thickness of the blade when setting up the cuts. Set the blade so it cuts on the waste side of the tenon layout line to keep from cutting into the tenon. If the tenon has four shoulders and cheeks, cut the narrow edges by clamping the workpiece on end to a miter gauge outfitted with a tall auxiliary fence. Once you've cut all the cheeks, trim away the remaining waste by cutting the shoulders with the workpiece facedown or on edge against the miter gauge. Be sure to reset the blade height carefully so you trim just to the saw kerfs cut for the cheeks. Use a stop-block clamped to the miter gauge to index the shoulder cuts.

How to Cut Tenons

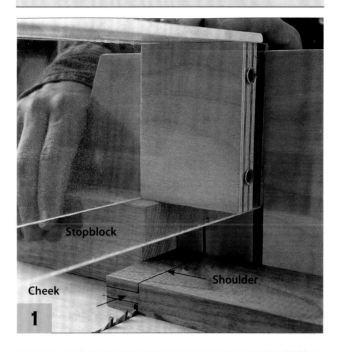

1 Lay out cut lines for the tenon on the end of your workpiece. Use a stopblock to index the shoulder cuts. For small workpieces, use a crosscut sled (page 91) if you have one.

2 Make additional crosscuts away from the shoulder of the tenon, removing waste wood. This part of the cut can get repetitive; take care to maintain your focus.

3 Smooth the tenon cheeks with a sharp wood chisel.

A dado blade will cut tenon cheeks and shoulders more quickly because it removes more material with each pass. Use the same multiple-pass technique as you would with a single blade to cut tenons with the workpiece flat on the saw table. Use a stopblock to establish the shoulder cuts, and make these cuts first. You can also stand the workpiece on end to cut a cheek and shoulder in one pass with a dado blade, but your depth of cut may be limited if you are using an 8"-diameter dado blade in your 10" table saw.

By all means clamp your workpiece securely to the tenoning jig if you are running a workpiece on end through a dado blade raised high above the saw table. Also slow down your feed rate. The dado blade will produce considerably more resistance on the workpiece as you make the cut because it is removing much more wood.

How to Cut a Haunched Tenon

When making heavy-duty frames, such as for building doors, extra twist resistance is always desirable. The haunched mortise-and-tenon is significantly more rigid than the blind mortise-and-tenon. Start cutting the tenons as with any other tenon: here, a dado blade is being used to "hog out" the waste wood.

Clamp the workpiece on edge to your miter gauge auxiliary fence. Trim the bottom of the tenon to size.

Cut the haunched portion of the tenon, usually starting about ¾" from the shoulder. Cut the mortise into the mating workpiece, using the haunched tenon as a template for outlining the cut.

How to Cut an Open Mortise-and-Tenon Joint

1

Clamp the workpiece to a tenoning jig and cut the tenon cheeks.

2

Trim off the waste to create the shoulders. Use your miter gauge and auxiliary fence to feed the workpiece into the blade, indexing the cuts with a stopblock.

3

Cut a through-mortise using the same setup you used to cut the tenon cheeks. The mortise should be equal in thickness and depth to the tenon, and centered on the workpiece.

4

Test the fit and adjust the tenon thickness, if necessary, by sanding. Apply glue to the mortise and assemble the joint. If the tenon is too long, plane or sand it flush after the glue sets.

FINGER JOINTS

Finger joints (sometimes called *box joints*) typically are made on a table saw with a dado-blade set and a jig—an auxiliary board screwed or clamped to the miter gauge. When joining parts of equal thickness, a finger joint is a good choice because it's strong and effective. Like a dovetail, the finger joint is visible after it's assembled—a plus if you like to show off your handiwork (and what woodworker doesn't?). Unlike dovetail pins, finger joint pins are straight, so it's an easier joint to make than a dovetail, although it's not as strong.

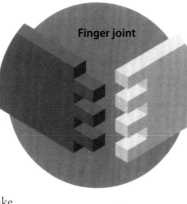

Finger joint

To make accurate finger joints, first rip-cut and crosscut the parts to size. Cut some test slots in waste pieces with the dado blade and check the fit of an actual workpiece in the slot. The workpiece should fit snugly without having to pound it in.

Finger Joints

Stress Resistance:

Tension: Good

Compression: Good

Shear: Good

Racking: Good

Common Uses:

Corners on boxes and chests

How to Cut Finger Joints

1

Install a dado-blade set and throat plate in your table saw. Set the cutting width of the dado blade to equal the thickness of the finger pins to be cut. Raise the blade set to cut the full depth of the pins. Clamp an auxiliary fence board to your table saw miter gauge. The board should be about 6" wide and at least 18" long. Make a pass of the auxiliary fence over the blade, then cut a strip of hardwood to use as a pin to fit in the slot, and glue it into the fence slot.

2

Reset the auxiliary fence by moving it a distance equal to the thickness of one pin to the outside edge of the dado blade. Reclamp or screw the fence to the miter gauge.

3

With the pin spacer inserted in the fence slot and the fence in position, butt the first workpiece against the strip and make the first pass. You can hold the workpiece in place by hand or clamp it to make the cut. After the workpiece and fence clear the blade, shut off the saw and back the workpiece off.

4

Reposition the workpiece by placing the slot you just cut over the pin space, then make the next cut. Continue in this manner until all the joints in that board are cut. Flip the board end-for-end and cut the fingers on the other end of the board the same way.

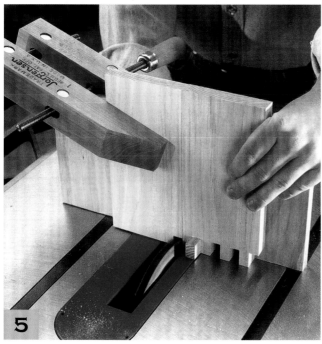

5

To cut the joints in the mating boards, fit the last notch you cut in the first piece over the pin, then butt the mating piece against the first piece, creating a one-notch offset. Make the first pass on the mating piece. Now remove the first piece, butt the notch in the mating piece against the pin, and make the second pass. Continue until all the joints are cut in one end, then flip the board end-for-end and repeat.

6

When all the joints are cut, the pieces are ready for assembly. Glue the joints and clamp them together with wood cauls offset from the joints to allow the joints to close.

SPLINED MITER JOINTS

Just as splines can help align and reinforce butt joints, they can also improve the strength and appearance of miter joints and bevel joints. When reinforcing miter joints, the spline is often cut from a contrasting wood to the members of the joint, which also adds a decorative accent to the joint. The spline can also be made of the same wood as the rest of the frame so it blends in with the joint.

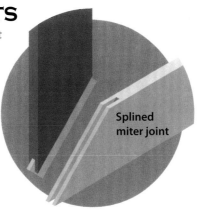

Splined miter joint

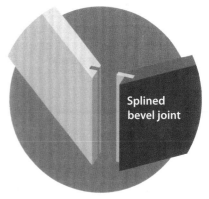

Splined bevel joint

When splines are used to reinforce a bevel joint, the advantages are entirely structural. The spline helps line up the joint corners during glue-up and reinforces the joint once the glue dries. The spline is visible from the top and bottom edges of the workpieces, but at least one set of these edges is typically covered by a structure above, such as a base cabinet over a beveled toe-kick, so the spline is essentially hidden.

Cutting Splined Miter Joints

Reinforcing a miter joint with splines involves cutting shallow slots into the mitered end of the workpiece in one of two ways. In the case of *splined miter joints*, a shallow kerf cut is made across the width of each workpiece, along the mitered end. When the joint is assembled, the spline fits evenly between the workpieces and across the full width of the joint. Consequently, the spline is exposed on both the inside and outside corners of the joint. When splines are used in ¾"-thick frame members, the spline is typically ⅛" thick and ¾" wide. The safest way to cut these splines is to clamp the workpiece on end against a simple scrap plywood jig supported from behind by a backer board fastened at a 45° angle to the plywood (see photo, page 135). Attach a tall auxiliary fence to the saw's rip fence. Set the blade height to ⅜" and center the blade on the width of the workpiece. Clamp the workpiece in the jig so the mitered end is flat against the saw table. Slide the jig along the fence to cut each spline slot.

A second option for splining a miter joint is to cut a spline slot through the outside corner of the miter joint, perpendicular to the 45° line formed by the joint parts; this is called a *feather joint*. You'll need to build a spline-cutting jig to cut these slots, since neither long edge of the mitered workpieces touches the saw table. (For plans on building a spline-cutting jig, see pages 158–159.)

Most spline-cutting jigs consist of a V-shaped cradle to support the mitered parts, with a backer board attached to the cradle to provide a clamping surface. Some splining jigs ride in the saw's miter slots, but most ride against the saw fence. It's a good idea to glue up the miters and allow the glue to dry before cutting the splines. This ensures that the frame parts stay aligned in the jig while you make the spline cut (but this isn't a necessity, especially if you clamp the workpieces in place firmly in the jig).

To cut the spline slot, raise the blade so it protrudes through the slot in the jig cradle. Measure the width across the mitered corner of the workpieces and set the blade height so it cuts one-half to two-thirds of the way across the corner of the miter joint. The spline slot should be centered on the thickness of the joint parts. Cut spline stock from matching or contrasting wood and glue it into the spline slot so the spline grain runs across the joint. Trim the spline flush to the outer edges of the mitered parts with a handsaw or router and flush-trimming bit.

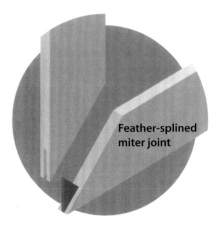

Feather-splined miter joint

Splined Miter and Bevel Joints

Stress Resistance:

Tension: Fair

Compression: Good

Shear: Good

Racking: Good

Common Uses:

Reinforcing miter joints on frames and bevel joints on carcases

Cutting Splined Bevel Joints

Cutting splined bevel joints involves first beveling the ends or edges of the joint parts at 45° so the workpieces form a 90° corner when they are assembled. To cut the spline slots, set the beveled edge or end of the workpiece flat on the saw table with the bevel facing down and against the rip fence. Set up the cut so the blade tilts into the workpiece. Adjust the fence until the blade intersects the workpiece near the inside corner of the beveled edge, and raise the blade to ⅜". Cut the spline slots with the workpiece held against the miter gauge, sliding the bevel tip along the rip fence. Cut and insert splines just as you would for splined miters or splined butt joints.

Cut spline slots in beveled edges by tilting the blade to 45° and guiding the workpiece through the blade with your miter gauge. Set up the cut by visually aligning the workpiece with the blade: for maximum holding power, the blade should enter the beveled edge very near the inside corner see Proper Spline Position in Miter Joints, below. Unplug the saw before setting up for the cut. Because you're not making a through cut, you can use the rip fence to help index the cut.

Proper Spline Position in Miter Joints

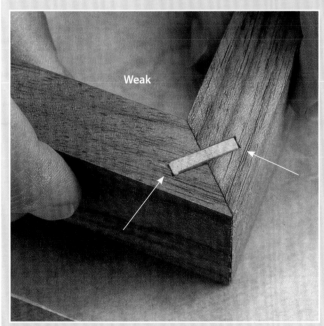

Weak

This spline location makes for a weak joint. The areas indicated by the arrows are too narrow, so the outside corner of the miter could break off.

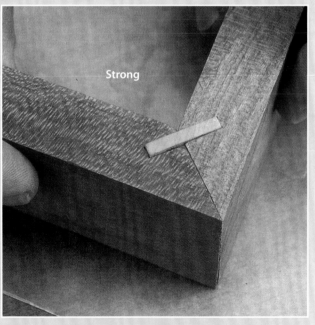

Strong

The optimal location for the spline is near the inside edge of the joint, leaving a thicker area above the spline.

How to Reinforce a Frame with Feather Splines

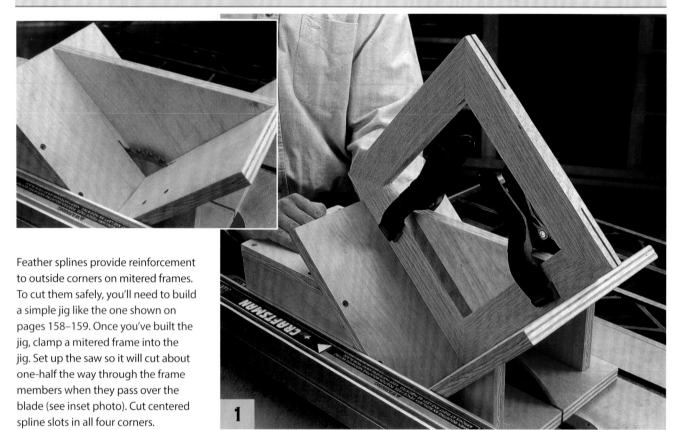

Feather splines provide reinforcement to outside corners on mitered frames. To cut them safely, you'll need to build a simple jig like the one shown on pages 158–159. Once you've built the jig, clamp a mitered frame into the jig. Set up the saw so it will cut about one-half the way through the frame members when they pass over the blade (see inset photo). Cut centered spline slots in all four corners.

1

2

Cut squares of stock the same thickness as the spline slots and glue the squares into the corner slots. After the glue has set, trim the feather splines so they're flush with the edges of the frame. Use a fine-toothed handsaw to trim the feather splines. Smooth out the joints between the splines and the frame with sandpaper.

Use Splines to Align and Reinforce Miter Joints

In some cases, splines are "retrofitted" into completed joints to add reinforcement (see previous page) or even to repair a corner. More often, though, they're included in the plan from the outset. Cutting spline slots into mitered boards can be tricky, but if you have a good, sturdy jig, it can be done with relative ease on a table saw. The jig shown here is really nothing more than a small hold-down clamp that's been secured to a square backer board. The clamp is positioned above a guide strip that's mounted on the same plywood backer at a 45° angle to the bottom edge. Simply clamp your mitered frame pieces against the guide strip and ride the jig along a tall auxiliary rip fence to cut the spline slots. Then glue a spline into each corner joint and trim after the glue has set.

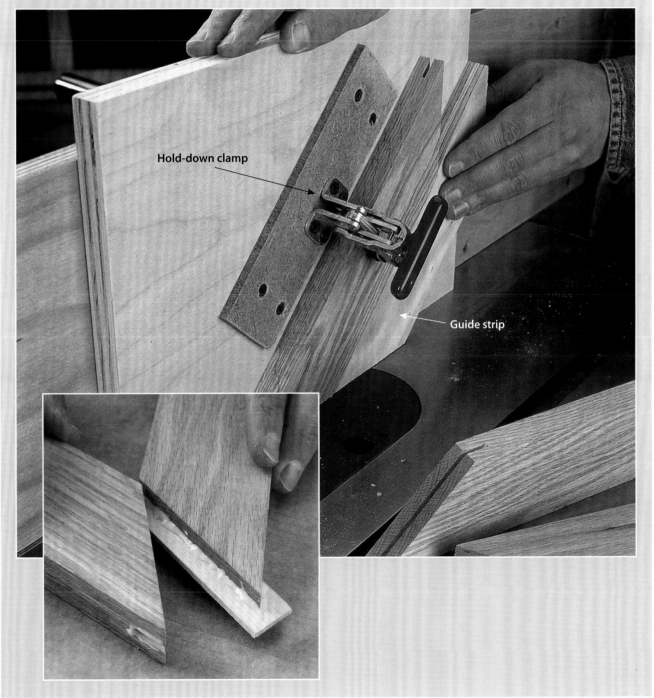

Hold-down clamp

Guide strip

SAWING ACCESSORIES

There are plenty of good reasons for modifying a table saw instead of replacing it. Perhaps your machine wasn't exactly top-of-the-line when you bought it. Although the saw may serve 90% of your needs, a couple of features or parts don't measure up. Maybe the rip fence won't stay parallel to the blade, so it seems like you are adjusting it every time you make a rip cut. It could also be the saw's miter gauge scale is too crude to set miter cuts accurately. Are the guard and splitter still mounted on your saw? If not, it's probably because these safety devices simply got in the way of your sawing, so you took them off for good. Most older saws,

regardless of price, suffered from poorly designed guards and splitters. Evaluate those aspects of your saw that suit your needs, then take a hard look at what falls short. A few improved parts may be all you need to perfect the tool you already own.

Accessorizing your saw could be the answer if your woodworking interests demand more from your saw than they once did. Aftermarket accessories could help lift your woodworking skills to a whole new level. If you cut loads of sheet goods, you might consider adding a permanent outfeed support system to your saw to make working with

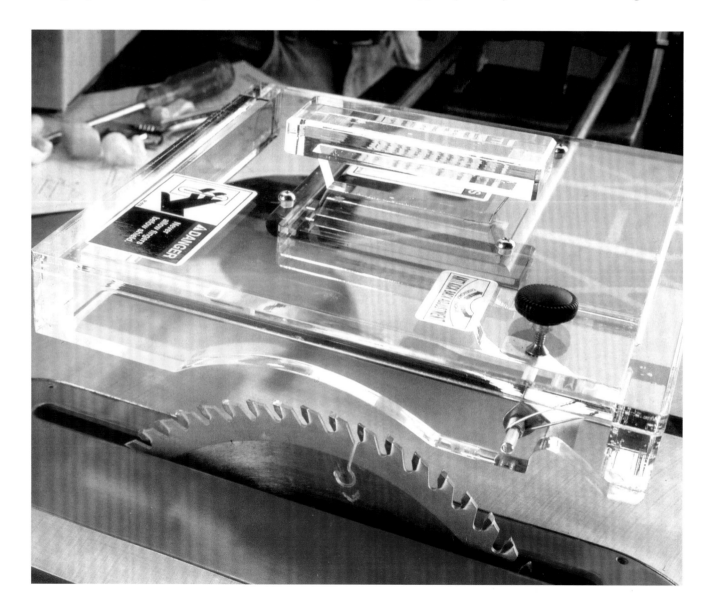

large sheets more manageable. If you find yourself regularly cutting mortise-and-tenon joints, a dedicated tenoning jig makes cutting joint parts a lot faster, easier, and more precise. A molding head and cutters might be just the solution you are looking for to create custom profiled molding without investing in a dedicated stationary shaper. In fact, you can even purchase sanding accessories that will convert your table saw into a sanding disk station.

Accessories can make your saw easier to move and cleaner to operate. Maybe you've grown tired of sneezing sawdust every time you spend an afternoon in the shop, and you've decided it's time to buy a dust collection system. You'll need to install a dust port on your saw to connect the saw to the collector. If you own a contractor's or cabinet saw, the chances are good that there's a dust port accessory available to fit it. If floor space is precious in your garage or basement shop, and your saw can't stay in one place at all times, a rolling saw base makes moving your saw less of a chore and much easier on your back.

The following pages offer a sampling of accessories to get you thinking about enhancing the performance and versatility of your saw. You may also find the chart below helpful for evaluating your accessory options. Eventually, you'll probably want to invest in a new saw, and maybe that's the best solution for you now. Only you can be the judge. However, spending a few hundred dollars to make a good saw better could be a more sensible option than investing several times that amount on a brand-new saw.

Accessory Options by Saw Category

Accessory	Jobsite saws	Contractor's saws	Cabinet saws
Fences		■	■
Guards		■	■
Removable splitters		■	■
Mitering attachments	■	■	■
Workpiece support	■	■	■
Dado blades/molding heads	■	■	■
Sanding accessories	■	■	■
Dust collection attachments	■	■	■
Rolling saw bases	■	■	■

■ Available for some saws ■ Available for most saws ■ Available for all saws

RIP FENCES

Saws made prior to 2000 often had substandard rip fences. Rip fences are critical to accurate, clean rip cuts. If the fence on your older saw won't stay parallel to the blade, slides roughly and binds along the fence rails even when the rails are clean, or if the fence is difficult to tune, you should consider upgrading your rip fence.

Fence systems have improved considerably in the last 20 years, but some saws still have lousy fences. Without a reliable and accurate rip fence, you'll never realize your saw's full cutting potential and you'll constantly fuss with fence adjustments.

If you are happy with your saw, but not your rip fence, take heart. Some manufacturers offer accessory rip fences to fit most popular full-sized saws. When combined with a side extension table, aftermarket rip fences can expand a saw's ripping capacity from around 2' to more than 4' in width (the extension table also provides additional support for crosscutting long workpieces). Unfortunately, the aftermarket saw fence industry still has little to offer owners of older jobsite saws. However, the rip fences on newer jobsite saws have vastly improved to meet the demands of the contractor market.

Adding an aftermarket fence will set you back a couple hundred dollars or more, not including the side extension table. The cost of the fence can even exceed what you paid for your saw in the first place. But the difference in performance and reliability will probably more than make up for your initial investment, and you'll wonder how you ever did without a better fence.

Generally, aftermarket rip fences are T-square style, where the fence clamps to a large hollow rail mounted to the front of the saw. The back of the fence rests on either the saw table or a metal ledge that bolts to the back of the saw. Fence bodies are made of extruded

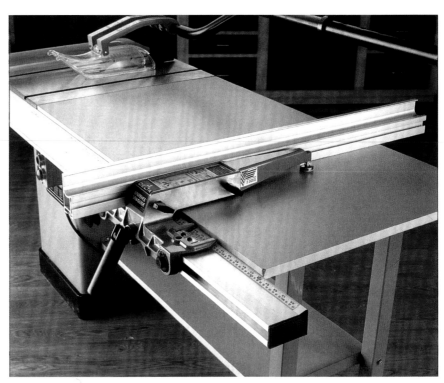

This T-square fence, built by Delta International Machinery Corporation, has a long aluminum fence body with a massive clamp assembly that attaches to the front rail. Larger clamps provide more support for the fence, which help keep the fence from deflecting during use. The fence body is long enough to suit saw tables of any depth.

Most aftermarket T-square fences resemble this one, built by the Biesemeyer Corporation. These fences can be purchased with a fence rail that spans the length of the saw table only or a longer version for use with a side extension table. Buy a fence that has an easy-to-read measurement scale along the front fence rail.

aluminum or steel and range in length from 3' to 5' to accommodate saw tables of varying depths. Since T-square fences clamp only to the front rail, fence bodies tend to be wider than they are tall to improve lengthwise stiffness.

Some T-square fences are faced on both sides with low-friction melamine plywood or polyethylene plastic to help minimize friction between the workpiece and the fence. Depending on the manufacturer, these laminated or plastic fence faces can be replaced if they are damaged.

On most fences, the clamp assembly that holds the fence to the front rail is massive. Some clamps are outfitted with ball bearings so the fence glides easily along the front rail. Other fence clamps simply slide along the rail.

In terms of retrofitting your saw, the installation process is fairly simple. Aftermarket fences may fit your saw exactly, if it is a popular model. Otherwise, you'll probably need to drill new holes in the saw table to accommodate the bolts that hold the fence system in place.

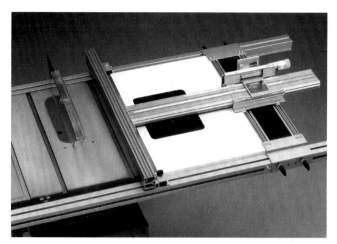

Incra Precision Tools manufactures a modified T-square style fence with the clamp mechanism located on the side of the saw table rather than the front. The fence slides in channels on rails mounted to the front and back of the saw. In this installation, the white extension wing doubles as a router table.

Features to Look For in a Rip Fence

T-square fence styles are virtually indistinguishable from one another at first glance, but the differences in quality are obvious when you look closely. The best way to choose a rip fence is to examine a few at a woodworking outlet or machinery dealer. Woodworking magazines will review fences occasionally, so read a few articles if testing actual rip fences isn't feasible where you live. When shopping for a new saw fence, here are some features to look for:

- The front clamp should lock solidly and squarely to the front rail. A fence clamp needs to hold the fence in place and stay parallel with the blade. Generally, the broader and more substantial the clamp base, the better.

- Fence movement along the full length of the front rail should be smooth. You should be able to unlock and slide the fence easily from one position to another, regardless of whether the fence rolls on bearings or slides against the rail.

- Every fence will have a measurement scale that stretches along the front rail. Be sure this scale is easy to read. The fence clamp will have some sort of cursor window for indexing the fence off of the blade. It should be prominent and adjustable.

- Adjustment features for squaring the fence to the miter slots and blade should be easy to operate. Be sure you can square the fence faces with the saw table as well.

- A micro-adjustment knob is a handy accessory, and it comes standard on some fences. This feature allows you to shift the fence in minute increments along the fence rail for precise settings.

The fence body that comes with Delta's Unifence system can be adjusted to any length across the saw table to function like a partial fence (see also page 78).

Adding attachments. Many saw operations will require adding an auxiliary (or sacrificial) wood fence to the rip fence. Look for a fence with holes or channels in the fence body that will make this installation easy.

GUARDS AND SPLITTERS/RIVING KNIVES

Guards and splitters or riving knives are your only mechanical lines of defense against injury from a table saw. Guards shield your hands and upper arms from exposure to the blade, and splitters or riving knives keep saw kerfs open to help minimize kickback. When guards and splitters were attached together to form a single piece of saw hardware, they were often clunky to use. They also were easy to knock out of alignment and limited the ability to see what was happening close to the cut. Even now, with safer riving knife assemblies, the guard must be removed for making non-through cuts, which places your hands and arms in full exposure to the blade.

For numerous reasons, woodworkers commonly removed the blade guard and splitter assemblies on older saws and stopped using them altogether. It might still be tempting to remove the guard and riving knife on newer saws, but this significantly breaches your safety from potential injury.

Guard Styles

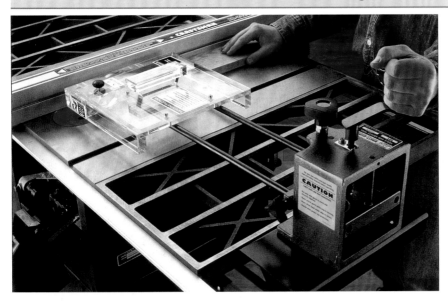

Crank-adjust blade guard. This guard assembly, made by HTC Products, Inc., uses a crank mechanism to raise and lower the acrylic blade shroud over the blade. Two metal posts attach the shroud to the crank housing. By loosening a couple of adjustment knobs, the shroud can be adjusted laterally across the saw table, a useful feature for those instances where the rip fence must be set close to the blade.

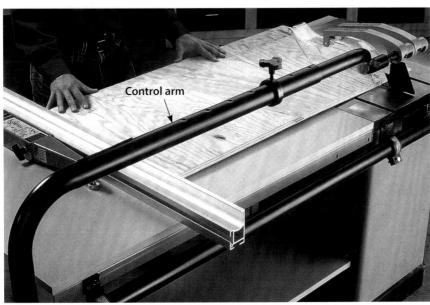

Delta's Uniguard system features a long control arm that attaches to the back of the saw and to a side extension table. The design creates plenty of clear space for crosscutting to the right of the blade. The control arm is made of two telescoping parts so the guard shroud can be adjusted laterally over the blade. This guard comes with a separate splitter, but not all aftermarket guards do. You may need to create your own splitter.

Flip-up convenience. One disadvantage of one-piece guard and splitter assemblies is you'll typically have to remove them for blade changes and maintenance. Delta's Uniguard shroud, on the other hand, conveniently flips up and out of the way for work around the blade.

Important Safety Features

Aftermarket blade guards operate independently of the splitter or riving knife, so they can remain in place for making non-through cuts, like dadoes. This setup will shield your hands from the blade, but it still won't protect you from kickback.

Delta's Uniguard shroud splits in two halves, so the shroud can straddle the rip fence, if necessary, to keep the blade covered.

Removable Splitters

Aftermarket blade guards will protect you from blade exposure, but unless they come with a splitter or riving knife, you're still vulnerable to kickback. The Biesemeyer Corporation manufactures a removable splitter with anti-kickback pawls that mounts beneath the throat plate and the cradle assembly in back of the blade. The splitter installs in a matter of minutes and can be popped in and out in even less time. To remove or replace the splitter, simply lift off the throat plate and pull on a spring-loaded knob at the base of the splitter assembly. The knob connects to a through-pin that holds the splitter in place.

Fortunately, a handful of aftermarket guards are available, and all of them are improvements over the one-piece guard and splitter design on older saws. Like aftermarket fences, companies that build replacement guards target the largest segment of the table saw market—full-sized contractor's and cabinet saws. There are no aftermarket blade guards for jobsite saws at time of writing.

Most aftermarket guards consist of a metal or plastic shroud suspended over the blade from a control arm or a pair of metal rods. The guard floats over the blade so it can work independently of the splitter or riving knife, which makes cutting with a dado blade much safer.

The control arm assemblies on aftermarket guards attach to the saw in several different ways. Some fasten to the back of the saw cabinet with brackets. Others mount to the end of a side extension table on the saw, to the floor, or even to the ceiling above the saw. Most control arms have telescoping parts, so the arm can be shortened or lengthened to reposition the guard shroud laterally over the blade. This adjustment feature is important, particularly when a cut requires you to set the rip fence close to the blade. The goal is to keep a guard in place, no matter how close the rip fence gets to the blade. On a standard one-piece guard and splitter, you'd have to remove the guard for rip cuts narrower than about ½", in order to squeeze a push stick in between the rip fence and the blade. Removing the guard is always an unsafe practice.

Blade shrouds must be capable of being raised and lowered to accommodate workpieces of different thicknesses. Some aftermarket guards are counterweighted, so the shroud rides up and over the workpiece, then drops automatically back into place against the saw table when the workpiece leaves the blade. One guard style shown on the previous page requires the operator to raise and lower the shroud with a hand crank each time the workpiece thickness changes. A third guard shroud style splits into two pieces, so each half of the shroud works independently of the other. It's a sensible design because the shroud keeps the blade shielded, no matter how close the rip fence gets to the blade.

Features to Consider When Buying an Aftermarket Blade Guard:

- **Lateral adjustment.** Be sure the control arm has a lateral adjustment feature so the shroud can move as the rip fence gets closer to the blade.
- **Crosscutting capacity.** Evaluate how much crosscutting capacity you need to the left and right of the blade. Control arms that mount to the saw or a side extension table will limit your clear space on one side of the blade or the other. If you plan to attach a side extension table to your saw, clear space shouldn't be a problem.
- **Mobility.** If your saw needs to be mobile, you'll want a guard system that mounts directly to the saw, rather than to the floor or ceiling. Buy a guard that disassembles and reassembles easily if your saw gets transported regularly in a vehicle.
- **Splitter or Riving Knife.** If the guard system you have in mind does not come with a splitter or riving knife, make sure your brand of saw will accept one of these devices. Otherwise, you should make your own splitter or riving knife from a piece of metal or rigid plastic. A guard alone only shields you from exposure to the blade; it won't protect you from kickback.
- **ON/OFF switch.** The control arms of some guards come equipped with an ON/OFF switch to make shutting down the saw possible without reaching underneath the saw table. Although this isn't a necessary feature to have, it is a handy convenience item and gives you an added measure of safety if you should need to shut the saw off in a hurry.

Tenoning Jig

If you cut tenons only occasionally, it's probably worth your time to build an inexpensive tenoning jig. Shop-made tenoning jigs are easy to build (see pages 162–163), but you've got to construct them carefully. The face of the jig that supports the work must be absolutely square to the saw table, otherwise you won't be able to make the tenon cuts straight relative to the workpiece.

On the other hand, if you cut tenons for virtually every project you build, a commercial tenoning jig like the one shown here is probably a worthwhile investment. Although relatively pricey, these heavyweight jigs are made of cast iron and steel, so they hold workpieces securely and squarely as you feed them into the blade. A large C-clamp-style screw holds the work against a vertical plate on the jig. The jig base is mounted on a steel bar that fits into and slides along either miter slot.

Tenoning jigs are adjustable in every direction, to suit virtually any tenon-cutting setup. Some even have a tilting feature so you can cut angled tenons as well. One locking lever allows you to tune the jig so the vertical support is precisely square to the table. Another lever controls lateral adjustments for setting up tenon cheek cuts. Better-quality tenon jigs are outfitted with a micro-adjustment knob to dial in precise lateral measurements.

Commercial tenoning jig. If you cut plenty of tenons, you'll appreciate the accuracy and convenience of owning a dedicated tenoning jig. Better-quality models are fully adjustable, sturdy, and extremely precise.

MITERING ATTACHMENTS

Cutting accurate miters is the bane of many woodworkers because it's tricky to do accurately. Miter angles that are off by even a degree can throw the joints of a picture frame or edge-banded corner out of whack, leaving an unsightly gap. The trouble is, when you are setting an angle on a standard-issue miter gauge, a two-degree difference is almost imperceptible, especially on low-quality miter gauges. Yet this level of precision is what you'll need to cut dead-on miters. Another problem with making accurate miter cuts is that it is difficult to hold a workpiece steady against the miter gauge fence without the workpiece shifting during the cut.

Several mitering accessories are available to help improve the accuracy of miter cuts. A number of manufacturers offer precision miter gauges with bars that fit the miter slots on most saws. Aftermarket miter gauges feature heftier housings, ultra-precise protractor scales, and even auxiliary fence attachments. Some have preset positive holes along the back edge of the protractor housing. A spring-loaded pin snaps into these holes to lock the miter exactly at a number of common angles, including 15°, 22.5°, 30°, and 45°.

The Dubby Crosscut Sled, made by In-line Industries, looks like a modified shooting board (see photo below). The jig is composed of a melamine-covered sled-style base mounted on a miter slot bar. An adjustable hardwood fence attaches to the top of the jig with lock knobs, and a metal scale runs along the edge of the jig for setting the fence to precise miter angles. The scale is positioned far enough from the fence's pivot point that distances between each degree mark are far enough apart to be set exactly. Another benefit

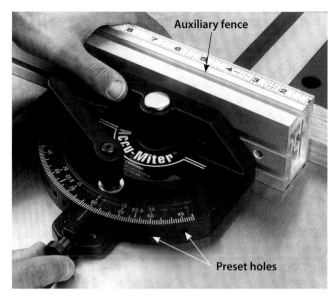

This precision miter gauge features a detailed protractor scale and preset holes for locking the fence at various angles. The gauge also comes with an auxiliary fence for added support.

to this mitering jig is that the workpiece can't shift during the cut because it is fixed to the jig.

An inexpensive route to improve your existing miter gauge is to fasten on a hold-down clamp. Hold-down miter gauge attachments work like quick-action woodworking clamps. A screw-type mechanism or pistol grip on the miter gauge tightens the workpiece against the miter gauge bar to hold it steady as you cut. The hold-down serves a safety function as well by keeping your hands off the workpiece and clear of the blade.

The Dubby mitering guide is equipped with a melamine-covered sled-style base that rides on a bar in the saw's miter slots for maximum workpiece support. A large miter scale, adjustable fence, and stopblocks make setting miter cuts easy and precise.

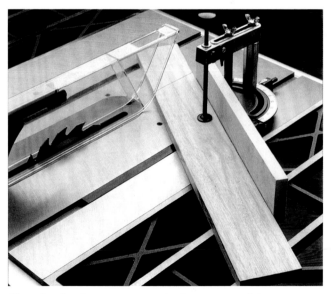

Clamping jigs fasten to your saw's miter gauge and clamp workpieces against the miter gauge bar to hold them securely. Some hold-downs have a knob to tighten the clamp, while others tighten by squeezing a pistol-grip handle.

MOLDING HEADS

Trim molding is a common design detail on many woodworking and remodeling projects. You can buy pre-milled molding in a variety of different profiles made from common wood types like pine and oak, but the cost can be steep. And eventually you'll run into situations where the molding profile you want simply isn't available, or you can't find it in the species of wood you are using in the rest of the project.

Milling your own molding isn't difficult, and you can save money in the long run by milling it yourself on the table saw using a *molding head*. A molding head is a heavy metal hub that spins on the saw arbor like a saw blade. Short steel blades, called *knives*, fit into two to four slots in the molding head. Knives are sold in matching sets, and each set is sharpened along an edge to form one of numerous different molding profiles including beads, coves, flutes, and grooves. Several different knife sets will come with the molding head.

Molding heads are somewhat controversial table saw accessories from a safety standpoint. Since you have to attach the knives manually, there's always the chance that you could forget to tighten a knife securely, and it could work loose during a cut and come flying off the cutter head at great speed. Also, since a molding head shapes rather than severs a workpiece, there is no saw kerf, so you can't use a splitter, riving knife, or anti-kickback pawls to safeguard yourself against kickback. The chances for kickback increase the more wood you "hog out" in a single pass.

You can use a molding head safely if you tighten the knives securely when you install them and take precautions to minimize your exposure to the knives when you make cuts. You can buy throat plates for molding heads, but the cutter opening is usually too wide to provide enough support close to the knives. A better alternative is to fabricate a zero-clearance throat plate or clamp a scrap piece of plywood over the throat plate area, and raise the molding head through it so that just the cutter profile protrudes through.

In most instances, you'll have to bury a portion of each knife inside the rip fence. To do this, attach a sacrificial wood fence to the rip fence just as you would when using a dado blade. Set the molding head below the saw table, position the auxiliary wood fence over the knife area, and slowly raise the molding head until the knives penetrate the wood fence to the height you need to mill the workpiece.

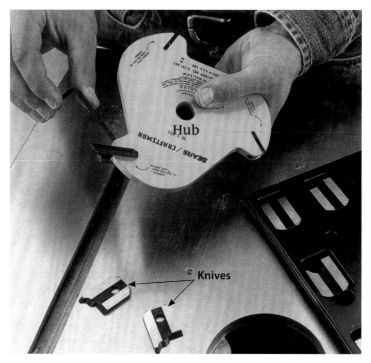

Molding heads consist of a heavy center hub with two to four slots for holding a set of profiled steel molding knives. Knives are available in a host of different shapes for cutting custom molding. Molding heads should only be used in full-sized saws powered by motors of 1½ h.p. or more.

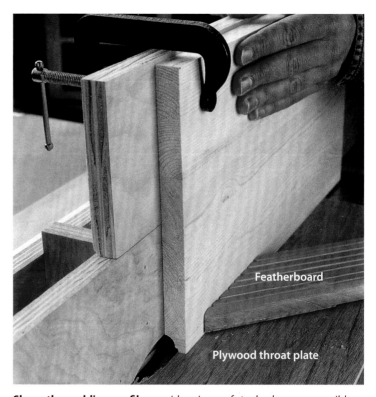

Shape the molding profile on wider pieces of stock whenever possible, then rip the narrow section of molding from the wider board. Be sure to use a zero-clearance throat plate or a piece of scrap plywood to shield as much of the molding head as possible. Featherboards will hold workpieces tight against the fence.

Cutting Molding Safely

In addition to using tight-fitting throat plates and wood fences, here are some more tips for using a molding head safely:

- Clamp featherboards beside the workpiece on the infeed side of the molding head to hold workpieces securely against the rip fence. Use a pushing tool to guide the workpiece through the cut.
- Cut moldings from wider pieces of stock, when possible, to keep your hands a safe distance from the cutters. Then rip the molding to width from the wider stock. For those situations where you must shape narrow stock, clamp a board on top of the workpiece and attach another board to the first to create a narrow "tunnel" over the cutting area (see photo, right). The tunnel will act as both a guard and a hold-down device around the knives. Stand clear of the back tunnel opening in the event a workpiece should shoot back out of the tunnel. When the back end of the workpiece enters the tunnel, use a long scrap piece of the same dimensions to push the workpiece through and finish the cut.
- Double-check that the knives are tight in the molding head before installing it on the saw and beginning a cutting operation.
- Regardless of the profile you are cutting, make a series of shallow passes, increasing the depth of cut by no more than ¹⁄₁₆" at a time.

Cutting Narrow Molding Safely

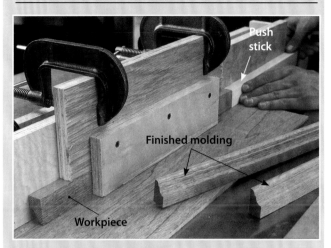

Molding "tunnel." In cases where you must cut molding on narrow workpieces, create a "tunnel" from scrap wood to support and guide workpieces as you feed them over the knives. Push sticks alone will not give you the holding power you'll need to control your work. Use a scrap push stick of the same dimensions as your workpieces to push the wood strips through the tunnel.

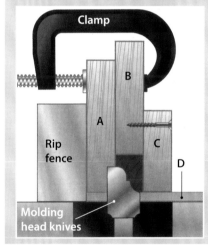

The molding "tunnel" consists of a sacrificial rip fence (A), a top support (B), side support (C), and either a zero-clearance throat plate or a piece of scrap plywood (D). The tunnel holds a workpiece (E) in place over the knives.

Sanding Disks

Convert your table saw into a sanding station by installing an aftermarket sanding accessory. The typical sanding kit includes a 10" metal plate and one or more circular sandpaper disks with adhesive backing. The plate mounts to the saw arbor just like a saw blade and fits through the throat plate opening. Depending on the manufacturer, some metal plates are flat on one face and tapered on the other. The tapered side makes it easy to sand chamfers, especially when you tilt the sanding plate. Sandpaper disks come in a variety of coarsenesses, from 60- to 180-grit, to suit most sanding tasks. Note that if you plan to do a lot of heavy disk sanding, buy a dedicated sanding station rather than using your table saw. Saw arbors are not engineered to withstand the kind of continual lateral stresses associated with sanding disks.

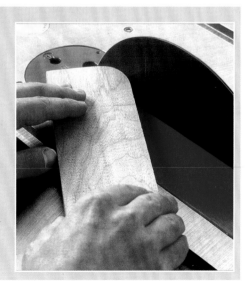

WORKPIECE SUPPORT

Any time you are making rip cuts or crosscuts on workpieces that project more than a few feet beyond the saw table, you'll need a support device alongside or in back of the saw to keep workpieces from tipping off the saw. You could set up one or more roller stands or slide a work table near the saw to support your work (see page 70), and these are fine solutions for occasional oversized cuts. However, if your woodworking involves frequent work with full-sized sheets of plywood or long lengths of stock, consider adding a dedicated extension table, outfeed roller platform, or sliding table to your saw instead.

Laminate-covered side extension tables are fairly common features on new saws that come with T-square rip fences, but you can mount an extension table to any saw. Typically, the extension table attaches to a steel base and replaces one of the saw's extension wings. The rip fence rails then attach to both the saw and the extension table. Some manufacturers offer laminated outfeed tables as well, for even more support. Another option for outfeed support is to bolt a roller platform to the back of the saw. These platforms are fashioned from a dozen or more ball-bearing tubular rollers (similar to those found on roller stands) in a metal frame. The platform is mounted on hinges so it can be tipped up and into position when needed or folded down flat against the saw to conserve floor space.

If your woodworking could benefit from a side extension table and your saw could benefit from better mitering control as well, consider adding a sliding table attachment. Sliding tables consist of a movable extension wing mounted to a metal frame that rides on ball bearings against the right- or left-hand side of the saw. A long fence attached to the table provides plenty of support behind your work and probably will come with a flip stop or two for making repetitive crosscuts. The fences on most sliding tables can be pivoted along a protractor scale for setting precise miter cuts.

Sliding tables mount to the side of the saw, replacing one of the saw's extension wings. The one shown here is made by Delta. Sliding tables offer superior workpiece support for crosscutting, because the whole table glides past the blade. This way, the workpiece can't shift during a cut. A long fence on the back of the saw table may even pivot on some models.

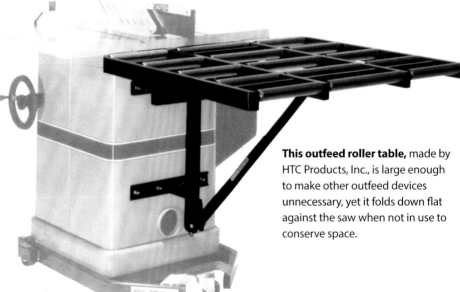

This outfeed roller table, made by HTC Products, Inc., is large enough to make other outfeed devices unnecessary, yet it folds down flat against the saw when not in use to conserve space.

Screw-on dust collection attachment. Cabinet saws that don't come with a round dust port connection can be retrofitted with attachments to make dust collection possible. This saw has an opening in the cabinet for a screw-on dust collection attachment available from the manufacturer.

Dust collection tray with hose port

To dust collector

Tray-style hose port. In order to connect a contractor's saw to a dust collector, you'll have to install a tray with a hose port to cover the bottom of the saw base. Some manufacturers offer these trays as aftermarket accessories for the saw. You can also build your own from a piece of plywood outfitted with a plastic or sheet metal hose port.

DUST COLLECTION ATTACHMENTS

You don't have to cut much wood in an enclosed space to realize how beneficial it is to connect the saw to a dust collection system. Table saws generate the lion's share of shop sawdust and chips, and quantities of this debris build up in hurry. If cleaner air isn't enough of an incentive, connecting your saw to a dust collector also makes shop cleanup a snap. The majority of saw waste will get sucked into a disposable bag on the collector. When the bag fills, just wind up the top and throw it away.

Many cabinet saws come with a 4"- or 5" round dust collection port built right into the saw base. It's sized to fit directly inside a flexible dust collection hose that attaches to the dust collector. Cabinet saws that don't come with a port will have some sort of removable panel or hole in the cabinet for installing a dust collection attachment with a hose port. These port attachments are inexpensive and available from the saw manufacturer. You also have the option to fabricate your own with a section of 4"-diameter stovepipe or ductwork and a piece of sheet metal.

It's more difficult to corral all the dust and chips produced from a contractor's saw because both the bottom and the back of the saw base are open. Some manufacturers sell a tray with a round dust port built in that fits over the bottom of the saw base. If a similar attachment isn't available for your saw, cover the bottom of the saw base with a piece of ¼" plywood fitted with a hose port. You may want to consider covering the back of your contractor's saw with plywood as well to cut down on dust. A back plate will improve the suction through the bottom plate hose port. You'll need to make cutouts for the drive belt and splitter/guard assembly. Be aware that this back plate will need to be removed when you tilt the saw arbor, since the entire motor assembly will tilt with the arbor and hit the back plate.

Most new jobsite saws have a port for attaching dust collection, but older saws may not. If your saw doesn't, you may want to consider designing a saw station with a dust collection chamber and port built in. Since the bottom of the saw base on jobsite saws is open, build a box-type saw stand beneath the saw to capture falling sawdust.

Regardless of which saw you own and how you attach your dust collection system to the tool, saw with a guard in place. A saw guard not only shields you from the blade, it also helps to direct dust and chips back down into the saw base rather than into the air.

ROLLING SAW BASES

Who doesn't want to have a dedicated shop space with enough room for each and every stationary tool? If you are not yet woodworking in your dream shop, it's likely you need to move tools around in order to use them, then move them out of the way to make room for other things that share your shop space, like cars or recreational equipment. A table saw is probably the first tool you'll get tired of dragging across the floor. The job gets much easier when you put your saw on wheels.

A number of manufacturers offer rolling bases custom-sized to fit full-sized saws. Extension tables are becoming so popular on saws these days that you can even buy one-piece roller stands custom fitted to fit the extension table.

Roller stands commonly consist of a steel frame outfitted with a number of wheels that fits underneath the saw base. One or more of the wheels typically lock into position with a foot-operated lever. When the wheels flip up, the saw drops down to the floor and is ready for use. Lower the wheels, and the base lifts up off the floor to roll the saw away.

If you'd rather not buy a prefabricated rolling base, you can buy kits with brackets and wheels to build a base that will fit any saw or stationary tool. To use these bases, you'll cut lengths of hardwood to span the distance between the corner brackets and essentially build the framework yourself.

One-piece steel rolling saw bases are available to fit virtually any brand of saw. You can even purchase bases that accommodate side extension tables, like the one mounted on this cabinet saw. Rolling bases turn an otherwise stationary tool into one that can be moved easily by one person.

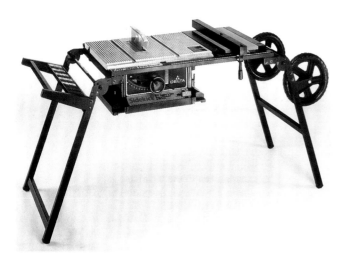

Portable saw base. Outfit your jobsite saw with a portable saw base, and you'll truly have a table saw that can go anywhere, indoors or out.

Saw base kit. A low-cost alternative to custom saw bases is to buy a rolling saw base kit and build it yourself. The kit comes with four metal corner brackets, wheels, and the foot pedal assembly that raises and lowers the saw. You supply hardwood struts to connect the brackets.

Converting a Saw Motor from 115 to 230 Volts

If your table saw is a contractor's or cabinet style, it is likely powered by an induction motor. Induction motors run at slower speeds and provide more constant horsepower under load than their lighter-weight counterparts, called *universal motors*. Induction motors are built for heavy, constant use. Universal motors are lighter-duty power plants usually found in portable tools like drills and circular saws, as well as most jobsite table saws.

One advantage to having an induction motor in your saw is that many manufacturers give you the option of rewiring the motor to run off 230-volt current rather than 115-volt household current. What this means for your saw is that it will run at half the amperage required when the motor is wired for 115 volts. Lower amperage means lower heat buildup—not more power. Over the long haul, less heat generated by the motor results in a longer motor life. The higher voltage supplied to the motor means it is less vulnerable to the significant voltage drops that occur when you plug your 115-volt saw into long extension cords.

It's easy to tell whether you can rewire the motor on your saw from one voltage to another. Motors that can be rewired will give you a wiring schematic right on the motor housing. You may also be able to find this information by removing the plate that covers the box where the power cord enters the motor. Your owner's manual should make this information clear as well.

Making the Switch

It really isn't difficult to convert your saw from 115- to 230-volt power, but of course your shop will need 230-volt service in order to plug the saw in. These higher voltage outlets actually have two "hot" wire leads rather than just one, and each of these hot wires supplies half the voltage the motor requires—or 115 volts each. A third wire in the 230-volt scheme serves as both a ground and a neutral lead. You can distinguish a 230-volt outlet from a 115; the top two slots for the plug prongs are oriented horizontally rather than vertically, so they look quite different from the typical three-prong household outlet. This configuration keeps you from unknowingly plugging a 115-volt appliance into a 230-volt outlet.

When you rewire an induction motor, what you'll effectively do is turn the power cord's white, "neutral" wire into a "hot" wire, which connects to a second unused hot wire inside the motor. The changeover is perfectly safe as long as you follow the instructions printed on the schematic drawing. A second change you'll need to make is to remove the 115-volt plug at the end of the motor power cord and replace it with a 230-volt plug. The power cord on motors designed for two voltages will be made of a wire gauge heavy enough to carry either 115- or 230-volt current safely. If you'd rather not splice on a new plug, you could replace the power cord altogether with one fitted with a 230-volt plug. Check with the manufacturer of your saw or buy the cord from an electric motor supplier in your area.

Induction motors that can be converted from 115 to 230 volts have a wiring schematic printed on the motor housing or on a removable cover to show you how to connect the wires. Follow the diagram carefully.

Making the conversion to higher voltage will require you to either splice on a new plug or replace the entire power cord. Notice that 230-volt plugs have two prongs that are turned 90° from a standard three-prong grounded plug.

DIY Workshop Essentials

CROSSCUT SLED

Crosscut sleds slide across the saw table from front to back and support workpieces that are too large to be guided effectively with the miter gauge. They are also helpful for making rip cuts and crosscuts on pieces that are too small to be held safely with a push stick. For more on crosscut sleds, see page 91.

The sled you see here has a pair of hardwood rails that ride in the miter slots on the saw table. The rails are fastened to a plywood base the same size as the saw table. The rails keep the sled traveling parallel to the blade as you push it across the saw. The base actually splits into two parts to make room for a blade slot, and plywood fences on the front and back of the jig hold the base parts in position. The fence on the infeed side of the saw also serves as a right-angle guide to support workpieces from behind, like a miter gauge.

For safety, an acrylic blade guard slides up and down in dado slots on both fences to protect you from blade exposure. A plywood blade tunnel mounted to the back of the sled keeps the blade shielded in cases where the blade must pass through the infeed fence.

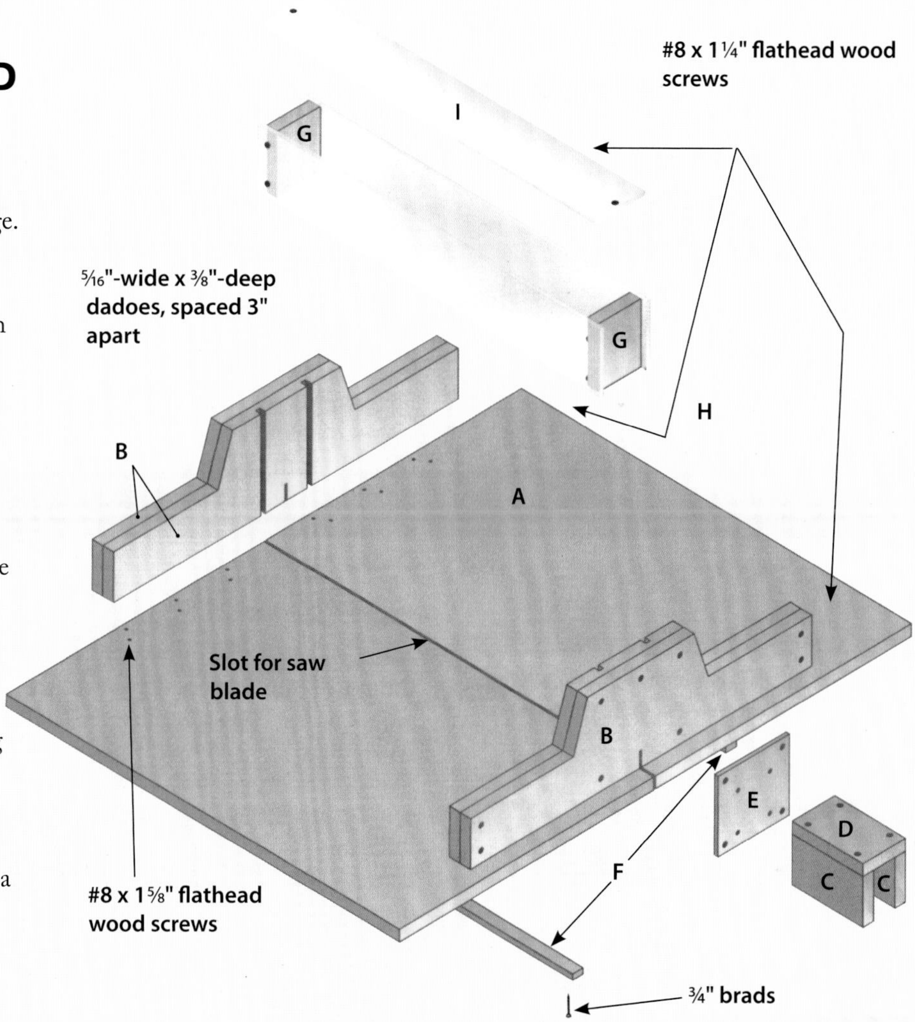

#8 x 1¼" flathead wood screws

⁵⁄₁₆"-wide x ⅜"-deep dadoes, spaced 3" apart

Slot for saw blade

#8 x 1⅝" flathead wood screws

¾" brads

Crosscut Sled Cutting List

	Part	No.	Size	Material
A.	Base	1	¾" x 27½" x 36"*	Plywood
B.	Fence	4	¾" x 24" x 6"	"
C.	Tunnel (side)	2	¾" x 3" x 5"	"
D.	Tunnel (top)	1	¾" x 3" x 5"	"
E.	Mounting plate	1	¼" x 4¾" x 5"	"
F.	Miter slot rail	2	⅜" x ²³⁄₃₂" x 27½"*	Hardwood
G.	Guard (end)	2	¾" x 4" x 3"	Plywood
H.	Guard (side)	2	¼" x 4" x 24⅛"	Acrylic
I.	Guard (top)	1	¼" x 3½" x 23⅜"	"

* Adjust part dimension to fit your saw

Crosscut sleds ensure that you'll make accurate, square cuts on workpieces that are too wide or long to be supported effectively by the miter gauge.

How to Build a Crosscut Sled

1 Cut the base and miter slot rails to size, according to the Cutting List, previous page. Size the base to match the dimensions of your saw table and the rails to match the width and depth of the miter slots. The rails should fit snugly in the miter slots and slide back and forth without binding.

2 Place the rails in the miter slots, then set the sled base on the saw table. Position the base so the ends and edges are flush with the saw table all around. Draw marks across the edges of the base and the ends of the rails on both the infeed and outfeed side of the saw table to reference the alignment of these parts **(see Photo A)**.

3 Remove the sled base and rails from the saw and attach the rails to the base. Spread wood glue on the edges of the rails that face the sled base, align the rails with the alignment marks you drew on the base in step 2, and fasten the rails to the base with ¾" brads. Attach one rail first, then set the sled on the saw table with the other rail in the miter slot. Check to make sure your alignment marks haven't changed for attaching the second rail as a result of attaching the first rail. It's crucial that the rails align perfectly with the miter slots once you've attached the rails to the base, or the sled won't fit properly into both miter slots at the same time.

4 Cut the four fence pieces to size. Fasten pairs of fence pieces together with glue and 1¼" wood screws to form two thick fences. We shaped the fences to be taller in the center in order to accommodate dado slots for the sliding blade guard but shorter on the ends to make it easier to hold workpieces against the fences. The fences are 3" tall on the ends and increase to 6" tall in the guard area. We made the guard area on each fence 8" long.

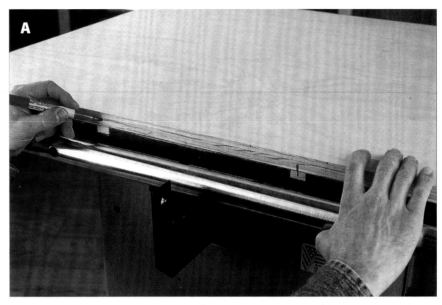

Photo A: Draw marks across the edges of the base and the ends of the rails on both the infeed and outfeed side of the saw table to reference the alignment of these parts. Line up these marks when you fasten the rails to the sled base.

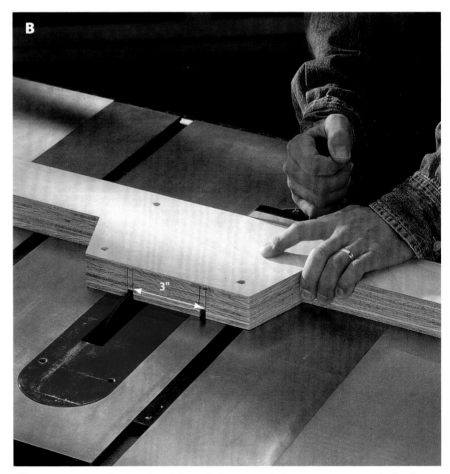

Photo B: Cut pairs of ⁵⁄₁₆"-wide, ⅜"-deep dadoes on each fence to serve as tracks for the blade guard. Space the slots 3" apart and center them on each fence.

5 Cut pairs of ⁵⁄₁₆"-wide, ⅜"-deep dadoes on one face of each fence to serve as tracks for the blade guard. Space the slots 3" apart and center them on each fence. Use a dado blade in the table saw to cut the slots (**see Photo B**).

6 Center one fence on one end of the jig base, perpendicular to the miter slot rails. Fasten the fence in place by driving 1⅝" screws through the base from beneath and into the fence. Set the jig on the saw, raise the blade so it cuts through the base, and cut a saw kerf through the fence and across the entire base (**see Photo C**).

7 Attach the second fence, using a carpenter's square to make sure it's perpendicular to the saw kerf (**see Photo D**). Center the fence on the saw kerf and position the outside face of this fence ¾" in from the edge of the sled base to serve as a lip for supporting the blade tunnel. Fasten the fence to the base with screws. Work carefully; it is critical that this fence be exactly square to the saw kerf, as it will serve as a right-angle miter-style guide for the sled.

8 Build the blade tunnel by cutting the sides and top piece to size and attaching them with screws. Fasten the tunnel to the mounting plate with screws, then attach the mounting plate to the outside face of the fence you attached in step 7.

9 Build the blade guard. Cut the two plywood guard end pieces to size. Cut the ¼"-thick acrylic guard sides and top using a fine-toothed plywood blade on the table saw. Sand the cut edges with emery paper. Drill countersunk pilot holes through the guard top and side pieces and into the guard ends. The guard ends should be flush with the ends of the top piece. The guard sides are longer and should overhang the top on either end to form runners that will fit in the fence dadoes. Assemble the guard with 1¼" wood screws (**see Photo E**).

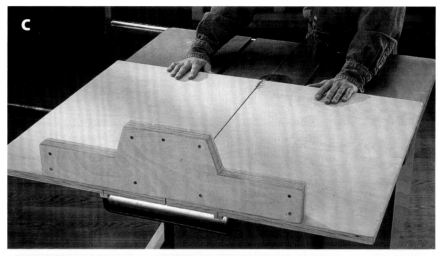

Photo C: Attach the outfeed-side fence to the sled base with screws. Raise the blade so it will cut through the base, and cut the blade slot across the base.

Photo D: Fasten the second fence to the sled, using a carpenter's square to make sure it's perpendicular to the saw kerf.

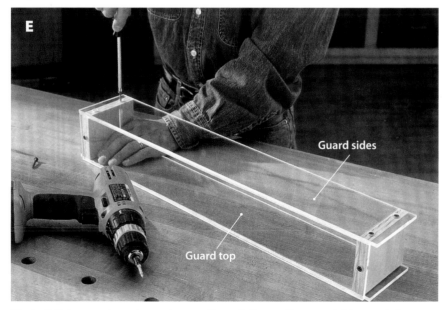

Photo E: Drive countersunk wood screws through the acrylic top and side guard pieces to attach them to the guard ends. The guard sides will extend past the guard ends to serve as runners for the guard to slide in the fence dadoes.

TALL AUXILIARY FENCE

Tall fences provide extra support whenever you need to cut a wide or long board with the workpiece standing on edge or end against the rip fence. You could bolt a wide piece of scrap directly to your metal rip fence, but fastening it this way is inconvenient if you need to take the fence on and off frequently. Build this tall fence jig instead, and you'll be able to clamp and unclamp the jig to the rip fence to install or remove it.

Tall Fence Cutting List			
Part	No.	Size	Material
A. Fence	1	¾" x 36" x 10"*	Plywood
B. Clamp rail	1	¾" x 3" x 36"*	"
C. Base	1	¾" x 2½" x 36"*	"
D. Braces	3	¾" x 2½" x 9"	"

* Adjust part lengths to fit your saw

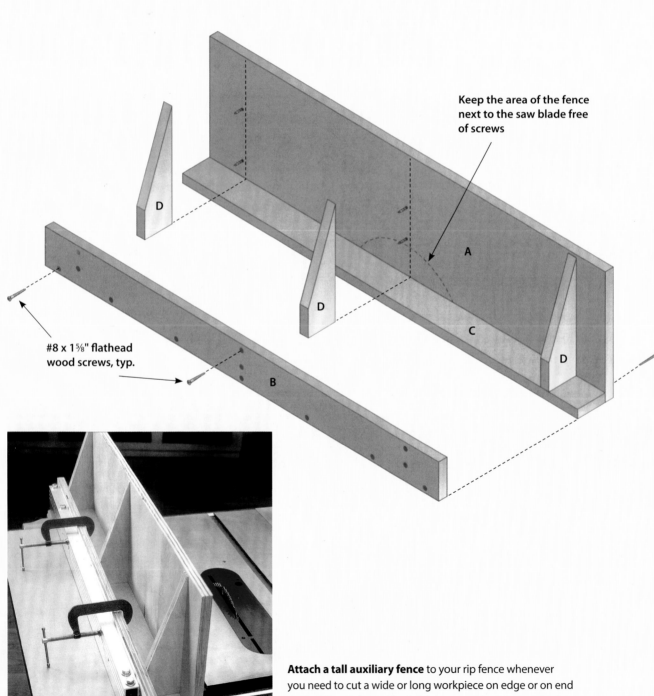

Keep the area of the fence next to the saw blade free of screws

A

D

D

D

C

B

#8 x 1⅝" flathead wood screws, typ.

Attach a tall auxiliary fence to your rip fence whenever you need to cut a wide or long workpiece on edge or on end against the rip fence. The tall fence provides more vertical bearing support for the workpiece. This fence design clamps quickly and easily to the rip fence.

1 Cut the fence and base parts to size. We used cabinet-grade plywood for the entire fence, but you could also use other dimensionally stable materials like MDF or particleboard to ensure that the fence parts will stay flat. Whatever sheet stock you choose to use, make sure the side of the fence that faces the blade is smooth. It will serve as the bearing surface for your workpieces, which need to slide smoothly past the blade when you cut them.

2 Fasten the fence to the base with 1⅜" flathead wood screws (**see Photo A**). Lay out locations for the screws, keeping them clear of the blade area. Drill countersunk pilot holes for the screws first so the screw heads will sit flush with or slightly below the fence face.

3 Cut the three braces to size. Be sure the bottom corners of the braces are square—they are responsible for keeping the tall fence square to the saw table. The braces will sit on top of the base and between the back face of the fence and the clamp rail. Attach the fence to the braces with screws. Position the center brace screws higher on the fence than the maximum blade height for your saw.

4 Rip-cut and crosscut the clamp rail, and then attach it to the fence base and braces with screws (**see Photo B**). The bottom edge of the clamp rail should be flush with the bottom of the fence base.

5 Ease sharp edges of the fence with sandpaper. To add durability and to create a slippery smooth bearing surface, cover the fence face with plastic laminate. You could also apply a few coats of gloss polyurethane varnish.

Photo A: Fasten the fence to the base with 1⅝" wood screws. Keep screws clear of the area on the fence where the blade will be when the fence is on the saw table.

Photo B: Attach the clamp rail with screws to the back square edges of the braces. Fasten the clamp rail to the fence base as well.

ADJUSTABLE TAPERING JIG

This adjustable tapering jig will allow you to cut accurate tapers with minimal setup time. Tapers reduce the proportions of workpieces like chair and table legs by cutting them lengthwise and at an angle to the blade.

Adjustable tapering jigs, like the design shown here, can be set and locked at a range of different angles to suit different taper dimensions. One leg of the jig rides against the rip fence. The other leg, outfitted with a short stopblock, supports the workpiece. For information on using a tapering jig, see pages 82–84.

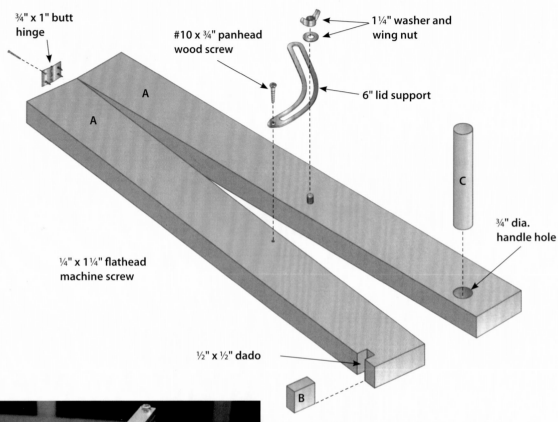

¾" x 1" butt hinge

#10 x ¾" panhead wood screw

1¼" washer and wing nut

6" lid support

A

A

C

¾" dia. handle hole

¼" x 1¼" flathead machine screw

½" x ½" dado

B

Tapering Jig Cutting List

Part	No.	Size	Material
A. Arm	2	¾" x 2½" x 22⅜"	Hardwood
B. Stopblock	1	½" x 1" x 1"	"
C. Handle	1	¾" x 4½"	Dowel

Additional hardware required:
- ¾" x 1" butt hinge with screws
- 6" metal lid support
- ¼" x 1¼" flathead machine screw, washer, wing nut

Tapering jigs allow you to cut a workpiece lengthwise and at an angle to the blade to trim tapers. This tapering jig project uses metal lid support hardware to lock the arms at a range of different taper angles.

1 Rip-cut and crosscut the two jig legs to size from flat, straight hardwood.

2 Cut a ½" x ½" dado into the edge of one leg, 1" from one end, to house the stopblock (**see Photo A**). Cut the stopblock to size and glue it into the dado. Glue alone is sufficient to hold the stopblock in place. Do not fasten the block in place with a screw. This way, if the blade should ever come into contact with the stopblock, the blade will not hit a fastener.

3 Cut the dowel handle to length from ¾" dowel, and install the dowel handle into the leg that will ride against the saw fence. Drill a ¾"-diameter guide hole through the leg, about 1" from one end and one edge. Glue the dowel in place.

4 Fasten the legs together with a ¾" x 1" butt hinge, attaching the hinge leaves to the ends of the legs opposite the stopblock and handle.

5 Install the metal lid support. Attach the lid support to the top face of the "stopblock" leg with a panhead screw. Place the screw 13⅜" from the hinge end of the jig and ½" from the inside edge of the leg. Drill a countersunk guide hole into the "handle" leg up from beneath the leg, 12¼" from the hinge end and ¾" from the inside edge. Insert the machine screw into the guide hole and through the slit in the lid support. Secure the machine screw to the lid support with a washer and wing nut (**see Photo B**).

Photo A: Cut a ½" x ½" dado into the edge of one leg, 1" from one end, to house the stopblock. The stopblock will be attached to the jig arm with glue only, so the dado provides more support.

Photo B: Attach the metal lid support to the jig arms, using the screw location information outlined in step 5.

Tip

Depending on the hardware you buy, you may have to first remove a short angle bracket attached to some lid support hardware. The bracket will not work on this tapering jig. To remove the bracket, drill through or grind off the rivet that fastens the bracket to the curved lid support brace.

SPLINE-CUTTING JIG

Spline-cutting jigs allow you to cut slots across the corners of miter and bevel joints in order to insert thin wooden splines. Splines strengthen and add a decorative element to a variety of wood joints, including picture frame, butt, and box joints.

Some spline cuts will require you to hold mating boards at a 45° angle off the saw table. For those kinds of cuts, the spline-cutting jig shown here employs a deep "V"-shaped cradle to support both boards and a backstop for clamping the workpieces securely while you saw. The cradle rests on a tunnel-type blade guard. The entire jig rides on a sled-style base and is guided across the blade by sliding the jig against the rip fence. For a demonstration on how to cut spline joints with this jig, see page 134.

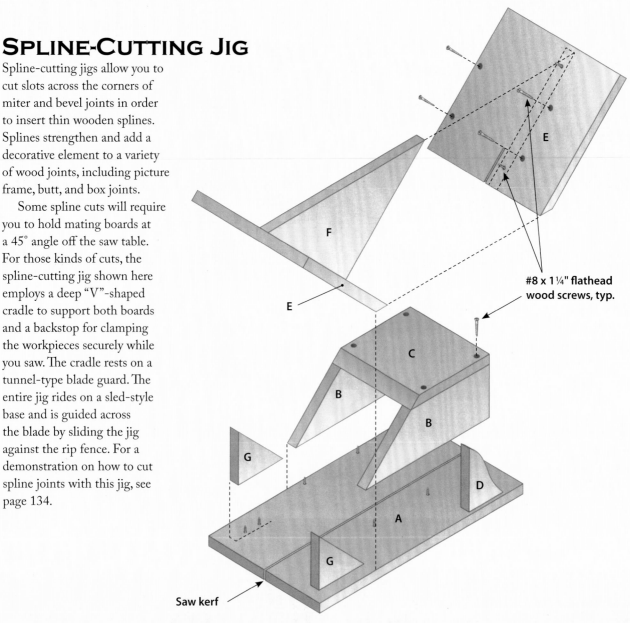

#8 x 1¼" flathead wood screws, typ.

Saw kerf

Spline-Cutting Jig Cutting List

	Part	No.	Size	Material
A.	Base	1	¾" x 10" x 18"	Plywood
B.	Tunnel (sides)	2	¾" x 4" x 11"	"
C.	Tunnel (top)	1	¾" x 7" x 7"	"
D.	Handle	1	¾" x 3" x 4¾"	"
E.	Cradle	2	¾" x 10" x 10"	"
F.	Backstop	1	¾" x 9⅞" x 9⅞"	"
G.	Braces	2	¾" x 4" x 4"	"

Some spline cuts will require you to hold workpieces at a 45° angle off the saw table, such as when making "feather" spline joints for picture frames (see inset photo, right). This spline-cutting jig provides a large 90° cradle for clamping workpieces in position. The jig rides along the rip fence, so the blade passes through the jig base and cradle to cut a spline slot in the workpieces.

1 Rip-cut and crosscut the plywood base and three tunnel parts to size. Miter-cut one end of both tunnel sides to 45°. Bevel-cut one edge of the tunnel top to 45°.

2 Attach the tunnel sides to the top of the base with 1¼" flathead wood screws. Position one tunnel side so it is flush with a base corner and aligned with the long edge of the base. The mitered edge of the tunnel side piece should face up. Fasten the second tunnel side to the base so the outside faces of the tunnel sides are 7" apart and the ends of the tunnel parts align. Fasten the tunnel top to the sides with the bevel edge facing up (**see Photo A**).

3 Rip-cut and crosscut the backstop and two cradle pieces to size. Cut the square backstop in half diagonally to form two triangles (you'll only need one triangle to serve as the backstop). Bevel-cut one end of each cradle piece to 45°. Set the backstop between the cradle pieces, and arrange the beveled ends of the cradle to form a short, flat surface at the bottom of the cradle assembly. Center the backstop along the width of the cradle. Attach the cradle pieces to the backstop with screws (**see Photo B**).

4 Set the cradle assembly in place on the jig base, resting the back of the cradle against the blade tunnel. Align the edges of the cradle with the edges of the base. Drive countersunk 1¼" screws through the cradle and into the blade tunnel sides.

5 Cut a 4" x 4" plywood square in half diagonally to form the two triangular braces. Attach the braces to the back face of the cradle, opposite the blade tunnel. Attach the braces to the jig base as well (**see Photo C**).

Tip

When you use the jig for the first time, set the jig in place against the rip fence. Measure the distance from the slot location you'll cut on your workpiece to the rip fence, and use this as the index for cutting a saw kerf across the jig base and cradle.

6 Cut and fasten the handle to the jig base next to the blade tunnel.

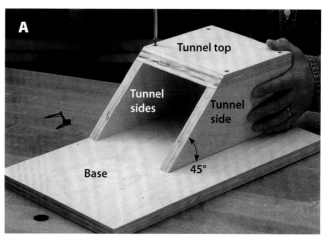

Photo A: Attach the tunnel sides to the base with 1¼" flathead wood screws. Position one tunnel side so it is flush with a base corner and aligned with the long edge of the base. Fasten the tunnel top to the sides with the bevel edge facing up.

Photo B: Set the back stop between the cradle pieces. Arrange the beveled ends of the cradle to form a short, flat surface at the bottom of the cradle assembly. Attach the parts with screws.

Photo C: Drive countersunk wood screws through the cradle and into the tunnel sides to attach the cradle to the tunnel sides. Attach the braces to the cradle and base with screws.

OUTFEED TABLE

Outfeed tables are a more stable alternative to roller stands for supporting long boards when you cut them on the table saw. The problem with most outfeed tables is they don't fold up or come apart for easy storage, so they take up valuable shop floor space.

If you have a small shop where space is at a premium, you'll appreciate the outfeed table shown here. The jig base mounts in the vise of a portable workbench or can be clamped to a sawhorse. The base is not permanently attached between the support pieces, so the tabletop and base slide apart for easy storage. Plus, since the jig clamps into a workbench or to a sawhorse, you can adjust the outfeed table height to suit any saw by raising or lowering the jig base in the clamps.

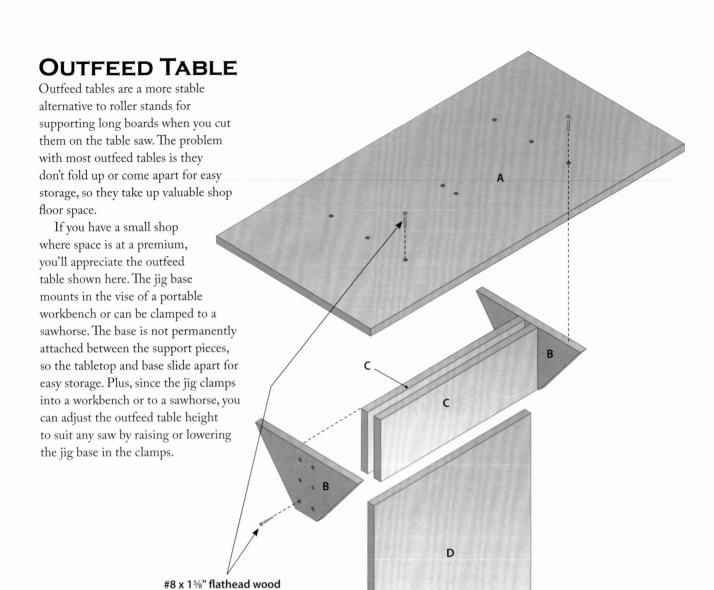

#8 x 1⅝" flathead wood screws, typ.

Outfeed Table Cutting List

Part	No.	Size	Material
A. Tabletop	1	¾" x 18" x 36"	Plywood
B. Support	2	¾" x 12" x 6"	"
C. Side	2	¾" x 6" x 18"	"
D. Base	1	¾" x 18" x 18"	"

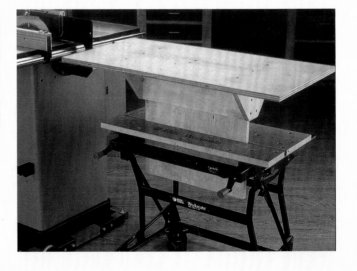

Outfeed table provide extra support behind or alongside the saw when cutting long or wide boards. This outfeed table can be clamped to a portable workbench or even to a sawhorse. The whole table can be adjusted up or down to match the table height of any saw.

1 Rip-cut and crosscut the plywood sides and supports to size. The supports are 12" wide at the top but slope to 2¼" wide at the bottom. Make the angled cuts on the supports with a power miter saw, jigsaw, or band saw.

2 Fasten the supports to the sides with 1⅝" wood screws. Slip ¾"-thick scrap spacers between the sides when you attach the parts to create a ¾"-wide slot for the base (**see Photo A**). Drive screws through the supports and into the ends of the sides. Remove the spacers to complete the subassembly.

3 Cut the tabletop and base to size. Center the subassembly you built in step 2 on the tabletop and trace its profile to serve as a drilling guide. Lay out positions for screws in the outline area, keeping screws clear of the recess between the sides. Drill countersunk pilot holes through the tabletop. Line up the subassembly beneath the tabletop and attach it to the tabletop with wood screws (**see Photo B**).

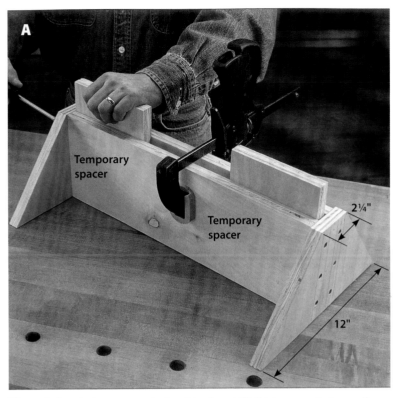

Photo A: Attach the supports to the sides. Insert ¾"-thick spacers between the sides to allow room for the base later.

Photo B: Drive countersunk 1⅝" wood screws down through the tabletop and into the supports and sides.

TENONING JIG

When you need to cut tenons with a workpiece standing on end on the saw table, a tenoning jig will provide the vertical support you'll need to make these cuts safely. A tenoning jig acts like a movable tall fence with a backstop attached to support workpieces from behind.

Tenoning jigs have other uses as well, including cutting through-mortises, rabbets, or dadoes in the ends of boards.

This tenoning jig is a snap to use. Simply clamp the workpiece against the tall support and the backstop so the end of the workpiece touches the saw table. Set the edge of the jig base (the edge closest to the handle) against the rip fence. Adjust the position of the rip fence and jig until the saw blade lines up with the cutting lines on the workpiece. To make your cuts, slide the jig along the rip fence and past the blade.

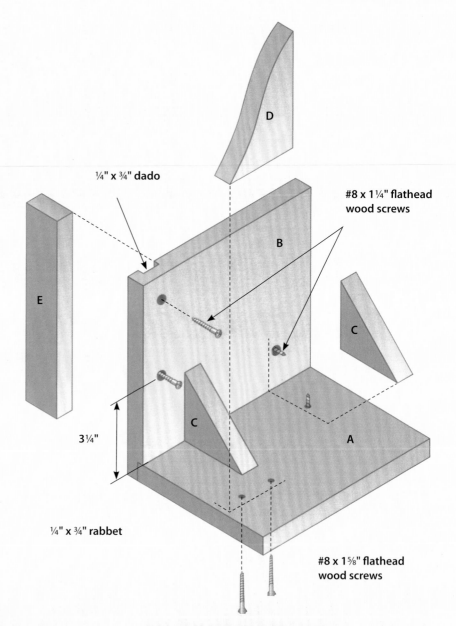

¼" x ¾" dado

#8 x 1¼" flathead wood screws

B

D

E

C

C

A

3¼"

¼" x ¾" rabbet

#8 x 1⅝" flathead wood screws

Tenoning Jig Cutting List			
Part	**No.**	**Size**	**Material**
A. Base	1	¾" x 6" x 8"	Plywood
B. Tall support	1	¾" x 8" x 8"	"
C. Brace	2	¾" x 2¾" x 2¾"	"
D. Handle	1	¾" x 3" x 4¾"	"
E. Backstop	1	¾" x 8" x 1½"	"

Tenoning jigs support tall, narrow workpieces on end for cutting tenons, rabbets, or through-mortises. To use the jig, clamp a board against the backstop so the end of the board rests against the saw table. Set the jig against the rip fence, and adjust the rip fence and jig until the blade aligns with the layout marks on your workpiece. Slide the jig along the rip fence to make the cut.

1 Rip-cut and crosscut the tall support to size. Cut a ¼"-deep, ¾"-wide rabbet into one of the faces of the support along one long edge. This rabbet will house the edge of the jig base.

2 Cut a ¼"-deep, ¾"-wide dado into the opposite face of the tall support, ½" in from the end (**see Photo A**). This dado should run perpendicular to the rabbet you cut in step 1. The backstop will fit into this dado. Since not all ¾" plywood is exactly ¾" thick, check the stock you plan to use for the backstop, and adjust the dado width as needed so the backstop will fit snugly.

3 Cut the base piece. Glue and clamp the tall support to the base. Check the fit of the parts to ensure that they are perpendicular.

4 Cut a 2¾" square of plywood in half diagonally to form the two braces. Fasten the braces to the back of the tall support and top of the base with 1¼" wood screws (**see Photo B**).

5 Drive countersunk screws through the tall support and into the backstop to fasten the parts (**see Photo C**). Position these screws at least 3¼" up from the jig base to keep them out of reach of the saw blade.

6 Cut and screw the handle to the base next to the braces.

Photo A: Cut a ¼"-deep, ¾"-wide rabbet into one of the faces of the support along one long edge. This rabbet will house the edge of the jig base. Cut a ¼"-deep, ¾"-wide dado into the opposite face of the tall support, ½" in from the end to fit the back stop.

Photo B: Attach the braces to the tall support and jig base with 1¼" wood screws.

Photo C: Cut the back stop to size and fasten it to the tall support with screws.

STRAIGHT-LINE RIPPING JIG

Crooked boards are commonplace, whether you buy lumber from a lumberyard, home center, or sawmill. Sometimes board edges are also bark-covered or otherwise uneven. Whatever the case, distorted board edges like these make them unsafe to rip on a table saw unless you can flatten one edge first. While minor crookedness is easy to correct on a jointer, more pronounced distortion is a bigger challenge. That's where this simple table saw jig can help. Designed to be fed along a table saw's rip fence, this 5'-long jig serves to immobilize crooked or defective-edged boards up to about 12" wide and 8' long. When a board is clamped on top of the jig with one edge protruding beyond the jig's base, the crooked or defective edge can safely be cut straight. Once straightened, it can serve as a reference edge for further processing.

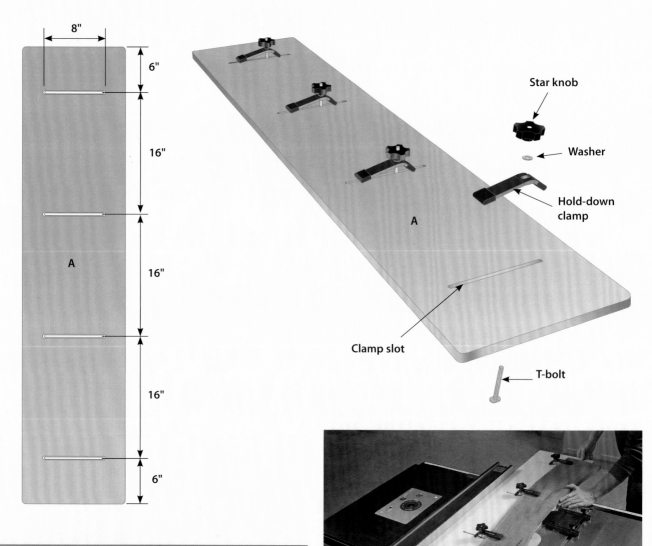

Tenoning Jig Cutting List			
Part	**No.**	**Size**	**Material**
A. Base	1	¾" x 14" x 60" (with ½"-radius rounded corners)	Plywood

Additional hardware required:
• Aluminum hold-down clamps

The straight-line ripping jig is designed to be fed against a table saw's rip fence. This jig serves to immobilize a board with two crooked edges. When the board is clamped on top of the jig with one edge protruding beyond the jig's base, it can be rip-cut straight for further processing.

Tips

Here are a few notes on the slots.

- Four ⁵⁄₁₆"-wide, 8"-long through slots accept ⁵⁄₁₆" x 3" T-bolts, aluminum hold-down clamps and star knobs

- Slots are spaced 6" or 16" on center

- Bottoms of through slots have a ⅛" x ½" x 8½" recess to house T-bolt heads and prevent them from rotating when the hold-down clamps are tightened or loosened

Photo A: File or sand the corners of the base before easing the top and bottom edges of the base with a ⅛"-radius piloted roundover bit.

1 Rip-cut and crosscut a jig base to size from ¾"-thick plywood.

2 Cut the corners of the base to ½" radii with a jigsaw, and file or sand them smooth. Then ease the top and bottom edges of the base with a ⅛"-radius piloted roundover bit in a handheld router to reduce the chances of splinters during use (**see Photo A**).

3 Draw a layout line across the base for each of the jig's four slotted holes. Mark lines 6" and 22" in from each end of the base. Then mark endpoints for the length of each 8"-long slot on the four layout lines (**see Photo B**).

4 Install a long ⁵⁄₁₆"-diameter straight bit in a mid-size handheld plunge router. Measure from the bit's center to the edge of the router's sub-base to determine the router's offset cutting distance (**see Photo C**).

Photo B: Mark two lines 6" and two lines 22" in from each end of the base, then mark the endpoints for the length of each 8"-long slot on these lines.

Photo C: For the router's offset cutting distance, measure from the bit's center to the edge of the router's sub-base.

5 Make a simple T-style straightedge to guide the router from a piece of wide scrap that's about 20" to 24" long and about 6" to 8" wide. Fasten a second scrap about 12" to 18" long underneath one end of the first scrap with screws to form a T shape (**see Photo D**).

6 Set the jig base on a spoil board that's at least as large as the base workpiece in order to protect your worksurface during the routing operation to follow. Position and clamp the straightedge you made in step 5. Its edge should be offset from the first slot layout line by the distance you determined in step 4, and its "T" crosspiece should be flush against the edge of the jig base workpiece (**see Photo E**).

7 Set the router into place with the edge of its sub-base against the straightedge and so the straightedge is oriented on the left side of the tool. Align the cutting edges of the router bit with the endpoint layout mark closest to you. Then start the router and plunge it into the base workpiece about ⅛". Push the router along the straightedge until the bit intersects the opposite endpoint of the layout line (**see Photo F**). Repeat this process several times, lowering the router's depth of cut for each pass until the bit cuts all the way through the base workpiece and slightly into the spoil board underneath.

Photo D: A T-style straightedge is easy to make. Simply fasten a 12"- to 18"-long scrap piece under another that's 20" to 24" long and 6" to 8" wide.

Photo E: The "T" crosspiece you made in step 5 should be flush against the edge of the jig base workpiece.

Photo F: Use the T-style straightedge to cut along the first layout line you marked in step 4.

8 Switch to a ⅝"-diameter straight bit in the router and repeat the routing process in step 6, but this time limit the depth of cut to just ⅛". This shallower routing pass will enable the heads of the T-bolts that secure the jig's hold-down clamps in place to recess into it. It also prevents them from spinning when the clamps are loosened or tightened (**see Photo G**).

9 Repeat steps 6 and 7 to rout the remaining three slots and bolt head recesses.

10 Insert a T-bolt through each slot in the jig base, and install a hold-down clamp, washer, and star knob onto the bolts. Make sure the T-bolt heads are engaged in the shallow recesses you made in step 7 (**see Photo H**).

11 To use the jig, clamp a crooked or defective-edged board on top of it and so one board edge extends completely beyond the edge of the jig (**see Photo I**). Secure the board to the jig with as many hold-down clamps as possible, tightened down along the board's opposite edge. Then set the saw's rip fence 14¼" from the blade. Install two featherboards in the saw's miter slot, one in front and one behind the blade. This keeps the ripping jig pressed against the rip fence during cutting. (Make sure the featherboards will not interfere with the board being cut when it passes over them.) Feed the jig along the rip fence as you would a normal rip cut to trim the protruding crooked or defective edge straight.

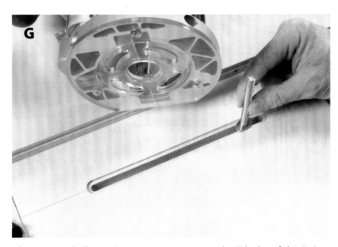

G

H

Photo G: A shallow ⅛" routing pass secures the T-bolts of the jig's hold-down clamps and prevents them from moving when the clamps are loosened or tightened.

Photo H: Install the clamps, washer, and star knobs onto the T-bolts to make sure everything is secure.

Photo I: Test your new straight-line ripping jig with a crooked or defective board before using it for a project.

INSTALLING A ROUTER LIFT

A table-mounted router is exceptionally useful for a variety of woodworking applications, including template-routing, general profiling, and joint-making operations. While freestanding dedicated router tables are common, if your table saw has an extension table made of plywood or other composite sheet materials such as particleboard, you can enhance its usefulness by installing a router lift into it. This way, the saw's rip fence can also be used for routing operations, and your table saw effectively becomes a two-machine tool.

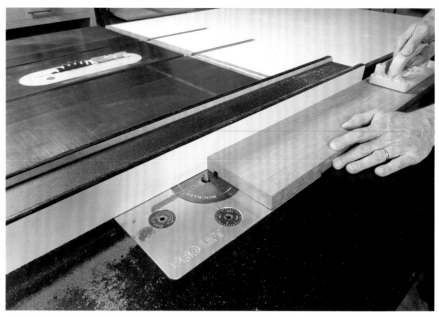

The router lift is great for making template-routing, general profiling, and joint-making jobs.

To install a router lift, you'll need a jigsaw as well as a freehand router. You'll also need a router mortising bit with a cutting radius that matches the radiused corners of the router lift's top plate. Typically, the radius will be ⅜", which will require a ¾"-diameter mortising bit. The bit should have a pilot bearing located on the shank above the cutter, and ideally, the bit's cutting length should be under 1".

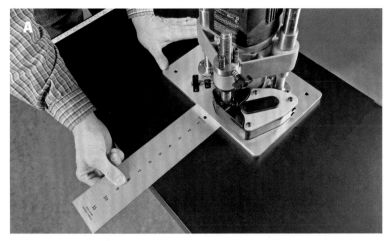

Photo A: For good workpiece support while routing wider material, leave at least 6" to 10" of table space in front of the lift.

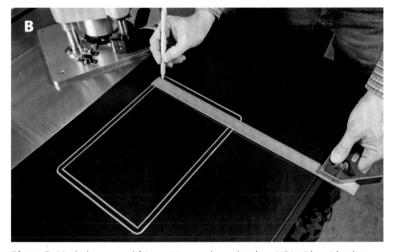

Photo B: Mark the router lift's position and another box ⅜" inside with a heavy pencil or marker.

1 Inspect your table saw's side table underneath to make sure it is free of metal framing members or other obstructions that could prevent it from being modified for router table use. Installing a router lift will require that the center portion of the saw's extension table be removed to fit the router lift's metal plate. It's also important that there be enough clearance below the saw table for the router motor to move freely up and down.

2 Position the router lift on the extension table so that it will be easy to operate for routing when standing beside the saw. Leave at least 6" to 10" of table space in front of the lift to allow for good workpiece support when routing wider material (**see Photo A**).

3 Trace around the router lift with a heavy pencil line or marker to mark its position. Then draw a second series of layout lines ⅜" inside the router lift perimeter layout line (**see Photo B**).

4 Use a ⅜"-diameter twist bit or Forstner bit to drill through the extension table at opposite corners of the interior layout line you just drew. These through holes provide clearance for a jigsaw blade in the next step (**see Photo C**).

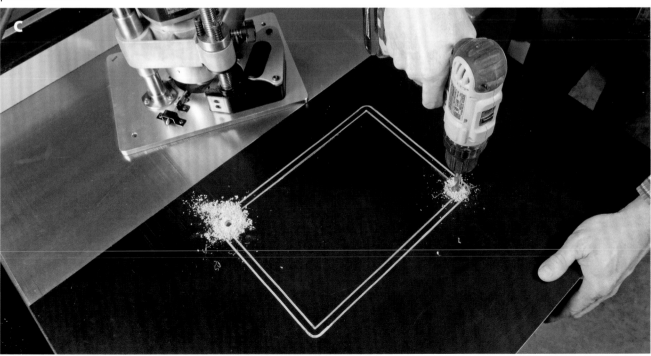

Photo C: Drill through opposite corners with a ⅜" diameter twist bit or Forstner bit.

5 Remove the interior waste area of the router lift cutout with a jigsaw, sawing along the interior layout lines (**see Photo D**).

6 Set the router lift back into position on the extension table so its edges are aligned with the original layout line. Use double-sided carpet tape to adhere four pieces of scrap material around the router plate (**see Photo E**). These scraps should be at least 3" wide, the same thickness, and form an uninterrupted "frame" around the router lift plate. They will serve as a support surface for a freehand router during the routing process that comes next.

7 Install the mortising bit in your handheld router. Set the router onto the support strip surface beside the hole in the extension table, and lower the bit so the rim of its top bearing will make contact with the inside edges of the support strips (**see Photo F**). Adjust the bit up or down as needed so the cutter will remove about ⅟₁₆" of material from the surface of the extension table in the first pass. If the bit's cutter is too long to set up the first pass in this fashion, you may need to install thicker support strips to accommodate the mortising bit size you have.

8 With the router bit clear of the contact surfaces, start the router and engage the bit gently into the inside edge of the router lift hole to initiate the cut. Then feed the router slowly around the hole, working clockwise (**see Photo G**). Keep the pilot bearing in contact with the support strip guide edges throughout the cut. The result of this cut will create a shallow rabbeted "ledge" around the inside perimeter of the router lift hole.

Photo D: Saw along the inner layout lines with a jigsaw.

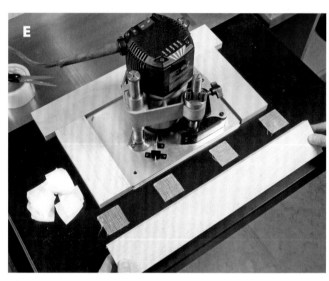

Photo E: Install four same-size scrap pieces around the router plate, using carpet tape squares to secure them.

Photo F: Make sure the mortising bit makes contact with the inside edges of the support strips.

Photo G: Work in a clockwise direction when feeding the router around the router lift hole.

9 Lower the bit about ⅛" for the second pass, and remove more material by making another routing pass.

10 The third routing pass will need to deepen the rabbeted ledge so it matches the exact thickness of the router lift plate. To set the bit's cutting depth accurately for this pass, take a scrap of the support strip material you used and hold it against the router lift plate to serve as a spacer. Set the bit's cutting depth to match the lift plate and spacer's combined thickness (**see Photo H**).

11 Carry out the third routing pass to bring the rabbeted ledge to its final depth. Pry off the support strips. Then set the router lift plate into place to see if it fits correctly (**see Photo I**).

12 The top face of the router lift plate needs to be level with the surrounding extension table surface. Check for this alignment with a long straightedge. Hold it diagonally across the lift plate in both directions as well as lengthwise and widthwise. If the lift plate stands slightly above the extension table, reinstall the support strips with tape, and carry out a fourth routing pass with the bit set slightly deeper to adjust the ledge depth. Alternately, if the plate is slightly deeper than the table, raise it to flush with strips of masking tape applied to the ledge to act as incremental spacers (**see Photo J**).

Photo H: Match the bit's cutting length to the combined thickness of the lift plate and spacer.

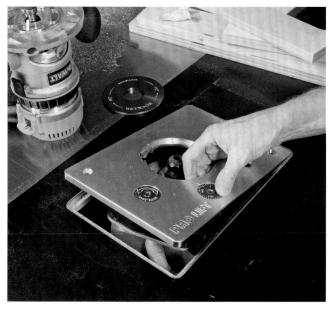

Photo I: Make sure the router lift plate fits.

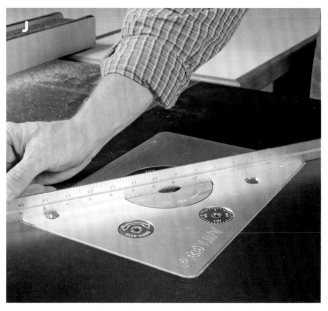

Photo J: Check that the face of the router lift is level with the surrounding table's surface with a long straightedge.

PART 2
Woodworking with Sheet Goods

INTRODUCTION

Sheet goods have become a staple for both the carpenter and the weekend do-it-yourselfer alike. We've all come to depend on plywood, particleboard, or MDF, but you may not think of these products as suitable for woodworking. In truth, the same advantages that make sheet goods terrific for building a roof deck or installing floor underlayment apply to woodworking. They're economical, you don't have to plane them to size, they're stable, and they can be topped off with any veneer or finish product you choose. Also,

pound-for-pound, some sheet goods are stronger than solid lumber.

As you browse through Part 2, we think you'll be impressed with the variety of sturdy and beautiful furnishings you can build for your home. You only need some quality sheet goods, a reliable table saw, basic woodworking tools . . . and the step-by-step instructions provided here.

A Sampling of Furnishings You Can Build with Sheet Goods

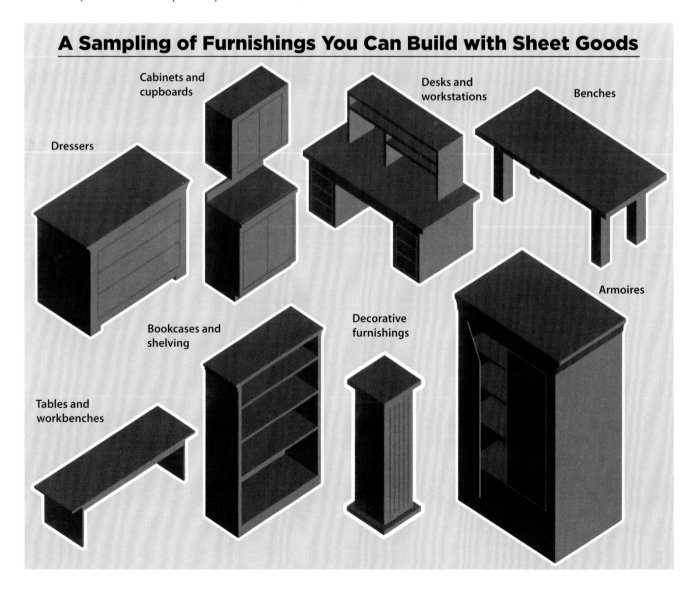

Cabinets and cupboards

Desks and workstations

Benches

Dressers

Armoires

Bookcases and shelving

Decorative furnishings

Tables and workbenches

TYPES OF WOOD

Plywood is manufactured in several thicknesses, using many different materials and processes to create the core, but ¾"-thick laminated veneer-core plywood with smooth hardwood veneer faces is the type used most frequently for woodworking projects.

PLYWOOD

Perhaps no building material has brought about greater changes in the way building tradesmen work than plywood. Originally developed for use as decking and subfloors, plywood quickly gained acceptance in the years immediately following World War II, where it was ideally suited for the mass-production demands that accompanied the postwar housing boom. Due to its ease of use and versatility, plywood soon found its way into lumberyards and building centers, where it was instantly popular with weekend do-it-yourselfers as well.

Initially, nearly all plywood was fashioned from sheets of softwood veneer—primarily pines and fir. By orienting the wood grain of each laminated sheet so adjacent sheets are perpendicular, the product was able to withstand greater stress than construction lumber of the same thickness. In addition, it was (and is) more dimensionally stable. The only real drawback to plywood was aesthetic: scarred with plugs and cracks on the outer faces and filled with voids on the edges, it simply wasn't much to look at. Since its arrival on the market, new grades of plywood with hardwood face veneer and a range of core options have essentially eliminated this drawback. Today, plywood is a favored working material for many designers and builders of fine furnishings, and it has become a standard for use in building custom cabinetry.

Most lumberyards stock furniture-grade plywood in several thicknesses and with a few face veneer options (pine, red oak, birch, and maple are the most common face veneers). Lumberyards and wood products distributors carry or can order plywood with dozens of additional veneer options. You can also find furniture-grade plywood at home centers, but the veneer options will be fewer than you'll find at a lumberyard. If you don't need a full 4' x 8' sheet, home centers will often stock 2' x 4' and 4' x 4' "handy panels."

Plywood Thickness Guide

Thickness (nominal/actual*)	Uses
¼" / ⁷⁄₃₂"	Back panels, drawer bottoms, frame-and-panel inserts
½" / ¹⁵⁄₃₂"	Small cabinet carcases, drawer sides
¾" / ²³⁄₃₂"	Structural components, large cabinet carcases, solid doors, drawer fronts, face frames, stretchers

*Actual thickness varies by type and grade: read grade stamp or take measurements to find individual panel thickness.

Plywood Core Types

The most common plywood core types are: A) Combination core (wood veneer and composition board) has smooth surfaces and holds screws well; (B) solid core (MDF) resists warping and has smooth surfaces, making it a good choice for tabletops in low-moisture areas; (C) veneer core has all-wood plies with alternating grain direction for light weight and high strength; (D) lumber core plywood is made by bonding face veneer to edge-glued strips of solid lumber; it is very rigid and good screw-holding capability, but has less dimensional stability than other core types.

Manufacturing Plywood

While plywood has only been a common building material since the 1950s, the concept of face-gluing multiple sheets of thin wood together to create a panel is centuries old. Naturally, the process has become more sophisticated as various mass-production manufacturing processes have arisen. The end result is an inexpensive, versatile building material that makes efficient use of our dwindling forest resources.

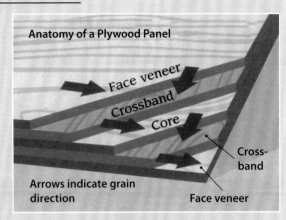
Anatomy of a Plywood Panel

Face veneer
Crossband
Core
Cross-band
Arrows indicate grain direction
Face veneer

The plywood manufacturing process starts with the selection and separation of suitable logs. The logs are peeled of bark, cut to length, then usually soaked or steamed before being mounted in a rotary cutter that slices them into sheets of veneer. The veneer sheets are dried, trimmed, sorted, plugged or patched if needed, glued, and then arranged by hand into multiple-ply "sandwiches" with cross-directional grain. The sandwiches of veneer are hot-pressed into plywood, some grades are sanded, and then each panel is individually graded and stamped.

Logs are peeled of bark, steamed, or soaked, then trimmed to just over 8' in length before they're thrown into a rotary veneer slicer. The slicer shaves the log into long, continuous veneer ribbons. Softwood veneer for the core and crossbands is sliced to a thickness of ³⁄₃₂" to ¼". Face veneers generally are sliced to about ¹⁄₃₂". The rotary slicer can produce up to 600 lineal feet of veneer per minute.

Sheets of green veneer are fed into a dryer for the curing process that reduces their moisture content to about 5%. The dried sheets are sorted and graded, then glued, heated to about 270°, and pressed into plywood panels under as much as 200 lbs. of pressure per square inch. Then the pressed panels are sanded, graded, and trimmed to size.

Special-Purpose Plywood

Bending plywood has a single-ply core (often lauan plywood) with face veneer (maple is shown as bending plywood, right); some types have multiple plies laminated with the same grain direction to allow flexing. Exterior-rated plywood is made with water-resistant glue and is used for cabinetry in high-moisture areas; choose panels with one B or better face.

Bending plywood

Exterior-rated plywood

Selecting Plywood

Choosing the right plywood for your project is an important task that can get complicated pretty quickly. In addition to the various core, thickness, and face veneer options, you'll also need to make a decision on the plywood *grade*. Despite the fact that trade-association sponsored plywood grading systems are clearly established, their application isn't always universal. This can complicate the task of ordering plywood and can lead to costly miscommunication. Basically, there are two grading systems in use today. The one most people are familiar with is administered by the APA (Engineered Wood Association, formerly the American Plywood Association). The APA grade stamps (see illustration, above) are found on sanded plywood, sheathing, and structural (called *performance-rated*) panels. Along with grading each face of the plywood by letter (A to D) or purpose (for example, *sheathing or sturdi-floor*), the APA performance-rated stamp lists other information such as exposure rating, maximum allowable span, type of wood used to make the plies, and the identification number of the mill where the panel was manufactured.

Many hardwood-veneer sanded plywood panels are graded by the Hardwood Plywood and Veneer Association (HPVA). (The grading organizations are trade groups whose stamps are used only by manufacturers who are members of the respective group.) The HPVA grading numbers are similar to those employed by the APA: they refer to a *face grade* (from A to E) and a *back grade* (from 1 to 4). Thus, a sheet of plywood that has a premium face (A) and a so-so back (3) would be referred to as A-3 by the HPVA (and AC by the APA).

When ordering some species of hardwood plywood, you'll find you have more than one option for veneer grain pattern. Most plywood has *rotary-cut* veneer (see illustration at right), which is the most economical type. But for surfaces that more closely resemble solid hardwood, look for sliced veneer faces that have strips of veneer, usually less than 1' wide, that are laid edge to edge. The strips may be *plain-sawn* (also called *face-sawn*) or *quartersawn*. If ordering from a mill, you can specify that the veneer strips be applied in the same order they're cut from the log.

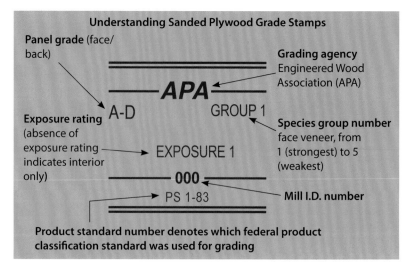

Understanding Sanded Plywood Grade Stamps

Panel grade (face/back)

Grading agency
Engineered Wood Association (APA)

APA

A-D   GROUP 1

Exposure rating (absence of exposure rating indicates interior only)

Species group number
face veneer, from 1 (strongest) to 5 (weakest)

EXPOSURE 1

000   Mill I.D. number

PS 1-83

Product standard number denotes which federal product classification standard was used for grading

Every sheet of plywood is stamped with grading information. On lower-grade panels, such as exterior sheathing, the stamp can be found in multiple locations on both faces. Panels with one better-grade face are stamped only on the back, and panels with two better-grade faces are stamped on the edges.

Face Grade Descriptions

N	Smooth surface "natural finish" veneer. Select, all heartwood or all sapwood. Free of open defects. Allows not more than six repairs, wood only, per 4 x 8 panel, made parallel to grain and well-matched for grain and color.
A	Smooth, paintable. Not more than 18 neatly made repairs, boat, sled, or router type, and parallel to grain, permitted. May be used for natural finish in less demanding applications. Synthetic repairs permitted.
B	Solid surface. Shims, circular repair plugs, and tight knots to 1" across grain permitted. Some minor splits permitted. Synthetic repairs permitted.
C plugged	Improved C veneer with splits limited to 1/8" width and knotholes and borer holes limited to 1/4" x 1/2". Admits some broken grain. Synthetic repairs plugged permitted.
C	Tight knots to 11/2". Knotholes to 1" across grain and some 11/2" if total width of knots and knotholes is within specified limits. Synthetic or wood repairs. Discoloration and sanding defects that do not impair strength permitted. Limited splits allowed. Stitching permitted.
D	Knots and knotholes to 21/2" width across grain and 1/2" larger within D specified limits. Limited splits are permitted. Stitching permitted. Limited to Interior and Exposure 1 or 2 panels.

Source: Engineered Wood Association

Plywood Veneer Grain Patterns

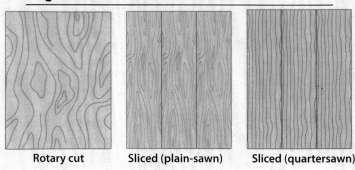

Rotary cut   Sliced (plain-sawn)   Sliced (quartersawn)

Common Face Veneers for Plywood

When using plywood to build woodworking projects, the species and quality of the face veneer are critically important to the final appearance of the project. For most projects, choose a product with an "A" grade hardwood veneer face on at least one side. Some of the more common wood species available for face veneers are shown here (without finish applied).

Clear pine. Species varies by geographic location; generally plain-sawn, sliced face veneer; cost is usually at least double the price of standard fir plywood with one A face.
Recommended finishes: Orange shellac or clear topcoat; light to medium wood stain; can also be used for decorative painting techniques, such as pickling and color-washing.

Mahogany. Both Phillipine and Honduran mahogany are available at better lumberyards (Phillipine mahogany is shown here); generally rotary-cut veneer; Phillipine varieties are inexpensive to moderate; Honduran is moderate to high.
Recommended finishes: Mahogany (red/purple) wood stain with hard, glossy topcoat.

Birch. A very versatile veneer that can be stained to resemble most major wood species; has subdued, but occasionally interesting grain figure; normally rotary-cut veneer; cost is inexpensive to moderate.
Recommended finishes: Wood stains of any color range with clear topcoat; frequently used for painted projects.

Cherry. If your goal is to build a woodworking project that doesn't look like it was made of plywood, cherry veneer plywood is an excellent choice. Usually plain-sawn, with sliced veneer strips; because color varies dramatically from sapwood to heartwood, hand-select panels after careful inspection. Cost is moderate to expensive.
Recommended finishes: Dark natural wood tone is better left unstained; protect with clear topcoating product.

Walnut. Another good choice for projects intended to look like they're made from solid hardwood. Most walnut plywood veneer is American black walnut, plain-sawn, and sliced. Cost is moderate to expensive.
Recommended finishes: Avoid dark stains; light walnut or mahogany stain can create richer wood tone. Apply clear topcoat.

Oak. Red oak shown, right is perhaps the most widely available hardwood plywood. Usually rotary cut, although sliced veneer is also available. Inexpensive to moderate. White oak plywood is costlier, but has a lovely golden cast and can be purchased with quarter-sawn veneer (a good choice for Mission-style furnishings).
Recommended finishes: Light to medium wood stain (oak or walnut) with clear topcoat.

Maple. A popular hardwood plywood that is relatively easy to locate. Light wood tones makes it a good choice for contemporary projects. Rotary cut in most cases. Cost is inexpensive to moderate.
Recommended finishes: Can be tricky to stain due to dense wood grain; topcoat only for a light appearance. A good choice for painted projects.

PARTICLEBOARD

Particleboard has developed a dubious reputation over the years, perhaps in part because it is used so extensively in the manufacture of low-grade, knockdown furniture that's mass-marketed through discount store chains. While it certainly has its limitations in woodworking pursuits, it also possesses several unique qualities that might make it just the right sheet goods choice for your next project—particularly if the project includes a counter or tabletop. Particleboard is very dimensionally stable (it stays flat and isn't likely to be affected by wood movement issues); it has a relatively smooth surface that provides a suitable substrate for plastic laminate; it comes in a very wide range of thicknesses and panel dimensions; and it is inexpensive. In addition, particleboard makes efficient use of wood chips that might otherwise be wasted, helping to conserve forest products.

Particleboard does have some drawbacks: it lacks stiffness and shear strength; it has poor screw-holding ability; it degrades quickly when exposed to moisture; it's too coarse in the core to be shaped effectively; it's heavy; and it's made with bonding agents that release potentially dangerous or irritating chemical vapors.

In some woodworking applications, particleboard can be used for purposes other than as a substrate for plastic laminate. It is used occasionally to build carcases and doors for light-duty cabinets in low-moisture areas, and it is often employed as an inexpensive shelving material (although it tends to sag as spans get longer).

For woodworking, particleboard is used almost exclusively as a substrate for plastic laminate.

Caution

Particleboard and MDF usually contain urea formaldehyde resins that continue to emit low levels of formaldehyde gas for at least six months as they cure. People with high sensitivity to chemical vapors should limit the number of composite panels added to a room at one time. Always wear a particle mask or respirator as required and provide adequate dust collection and ventilation when cutting or shaping these products.

Manufacturing Particleboard

Particleboard is made by using heat and pressure to fuse wood chips, shavings, resins, and bonding agents into dense, continuous panels (up to 2¼" thick) that are trimmed to standard panel dimensions. The wood ingredients on better-grade panels are distributed so the larger flakes and chips are at the core of the panel, and the finer shavings and sawdust are closer to the faces. This arrangement increases the stiffness and strength at the core, while creating surfaces that can be sanded to a reasonably smooth texture.

In some cases, admixtures are included in the particleboard recipe to create special-purpose wood panels. Among them are fire retardants and exterior-grade resins for moisture resistance.

Because the wood particle used to make particleboard can be as large as ¼", the finished product is generally coarser and less dense than other types of composite board, such as MDF (see next page).

Particleboard moves along a conveyor belt from the sander, past the blow bar (which checks for delamination), and to the stacker, which stacks the boards into units.

MDF

Medium-density fiberboard (MDF) is quickly becoming the industry standard for many woodworking and carpentry activities that involve sheet goods. It's similar to particleboard in constitution, as it's created by hot-pressing wood fibers and resins into dense sheets that are trimmed to size. But the main difference is that MDF is formed from wood and wood pulp that has been pulverized into individual wood fibers, rather than chips or flakes. The individual fibers fit together much more tightly than odd-shaped chips and flakes, creating a smooth, dense material that can be edge-shaped with a router. The difference between standard particleboard and MDF is a little like the difference between concrete and mortar: Mixed with large aggregate, concrete surfaces and edges tend to exhibit popouts and voids, whereas mortar mixed with fine sand is smooth on the surface, edges, and throughout.

The smoothness and density of MDF make it a good substrate choice for veneered projects: the rougher surface of particleboard and most plywoods do not bond as cleanly with thin wood veneer. You can even laminate layers of MDF to create table legs and other structural components that can be veneered or even painted. MDF is also increasing in popularity as a trim molding material.

Like particleboard, MDF does not have a great deal of stiffness, and it tends to sag if not adequately supported. It also tends to be fragile around the edges, swells, and degrades from constant moisture contact, and it does not hold screws well (although coarse-threaded wallboard screws can be reasonably effective). If you are using it to build structural project parts, try to use mechanical knockdown fasteners wherever possible, and don't skimp on the wood glue. MDF costs a little more than particleboard and is comparable in price to construction-grade plywood.

Medium-density fiberboard (MDF) is growing in popularity as a veneer substrate, paintable surface, and raw material for moldings.

Manufacturing MDF

Cooling wheels are used to cool composite panels (both MDF and particleboard) as they begin to dry and cure after hot-pressing. The cooling wheel lets air circulate across both faces of the panel before it is processed further.

Composite Board Thickness

Unlike most plywood, which is typically undersized in thickness by 1/32", composite boards such as particleboard and MDF are manufactured within .005" of their nominal thickness. When designing your woodworking project, be sure to use the actual thickness of sheet goods, especially on finer projects with little tolerance for error.

The process of forming MDF begins with the heating and grinding of white softwood until it is broken down to individual wood fibers no more than 1/8" in length. The fibers are coated with resin and (in some cases) additives and formed into a mat that's about 18" thick. The mat is hot-pressed at about 350°F, using 350 lbs. per square inch of pressure to compress it to finished thickness (from 1/4" to 1 1/4"). The continuous webs of MDF are cut to rough size, then loaded into cooling racks (see photo, above). Then the sheets are sanded lightly, cut to finished size, inspected, and given appropriate grade stamps.

MELAMINE BOARD

Melamine is a very popular sheet goods product for making shelving and building European-style cabinets. Generally fashioned with a particleboard core, most melamine boards have two faces that are surfaced with thermofused melamine. It offers the advantages of plastic laminate while saving you the trouble of applying the laminate yourself. Thicknesses range from ¼ to ¾". Stock colors at most lumberyards and building centers generally are limited to white, gray, almond, and sometimes black. Other colors may be available through special order. The panels are oversized by 1" in each dimension (a 4 x 8 sheet is actually 49" x 97") because the brittle melamine has a tendency to chip at the edges during transport. Plan on trimming fresh edges for your project.

The brittle surface and particleboard core are not friendly to screws or nails—the best fasteners to use when working with melamine board are knockdown styles (see page 185).

While most melamine board is surfaced on both faces, it is available with the melamine bonded to only one of the faces. The other face is left bare so you can apply laminate of a different color or style if you wish. Particleboard with one face veneered and the other surfaced with melamine, called *mela-quinella*, can be purchased for cabinetmaking (see photo, below).

Melamine board is faced at the factory with melamine laminate. The thermofusing process used to apply the melamine creates a much stronger bond than you can achieve with plastic laminate applied at the job site with contact cement (and saves you the trouble of doing the laminating). It also reduces the number of seams.

The Two Faces of Mela-quinella

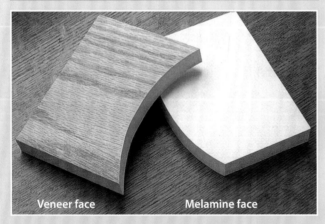

Veneer face   Melamine face

Designed mostly for cabinetmakers, mela-quinella is a hybrid between melamine and veneer-faced plywood. One face is finished with hardwood veneer, and the other face is coated with thermofused melamine intended to face inside a cabinet carcase or drawer. The veneer can be stained or top-coated to match the rest of the cabinets and face frames, while the melamine surfaces are ideal for cabinet interiors because they're bright and easy to wipe clean. The core is usually MDF or particleboard.

Cutting Melamine Board

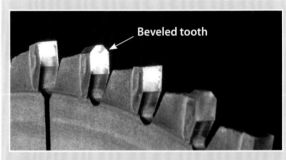

Beveled tooth

When cutting melamine board, as when cutting plastic laminate, the main goal is to avoid chipping the fragile laminate by using a blade that produces minimal tear-out. Blades designed specifically for cutting laminate are the best choice (see photo above), but if you'd rather not spend the money on a specialty blade, any plywood-cutting panel blade will do (these blades are distinguished by high tooth counts and minimal tooth set). When using a table saw, pass the material through the blade with the good (laminate) side facing up; when using a circular saw, cut with the good side face down.

MISCELLANEOUS SHEET GOODS

The universe of sheet goods is growing larger all the time, with new-and-improved and special-purpose products constantly being added to manufacturer inventories. Most of these products have little application in woodworking, but some, when used creatively, can provide an interesting or unique touch to your project. A few of the more useful types are discussed here.

Textured plywood. Designed to be used as exterior siding, textured pine plywood can be used as a back panel on open cupboards, or even as a cabinet carcase material on informal projects. The beadboard surfaced plywood shown here is readily available with grooves that are either 4" or 8" apart. Rough-sawn and brushed textures (with or without kerfs or channel grooves) can provide interesting surfaces for rustic or country-style projects.

Glue-up panels. Fashioned from strips of solid pine, these panels offer the benefits of solid wood: superior nail and screw holding, and ease of sanding and shaping. They do not require edge-taping or filling. The strips are finger-jointed together at the ends and edge-glued into ¾"-thick panels sold in a variety of sizes, ranging from 12" to 30" wide and 36" to 96" long. As a woodworking material, glue-up panels cost more per square inch than most plywoods, but are cheaper and more convenient to use than solid lumber. Higher-end versions are made from strips of select clear pine that are laminated together to increase thickness, finger-jointed to length, then edge-glued into prefabricated panels.

Tempered hardboard. Hardboard is used in woodworking mostly to make inexpensive panel parts that will not have high visibility: drawer bottoms and cabinet backs are two of the most common usages. It also makes good templates and patterns for project parts you may want to duplicate many times or make again in the future. Tempered hardboard (avoid the softer, non-tempered hardboard) can also be used as a bending plywood in areas that will not receive high stress or be exposed to moisture. It can be used as a veneer substrate in low-stress areas.

Perforated hardboard. Known commonly as *pegboard* or *perfboard,* perforated hardboard is simply hardboard (tempered or untempered) that is machined with ¼"-diameter holes every 1" on-center. For woodworking applications, perfboard may be used as a back panel material for cabinets that require ventilation, but its main use is as drilling guide material for locating shelf-pin holes in cabinet sides and standards.

Working with Sheet Goods

Handling and Storage

Sheet goods are awkward to handle, especially if you're trying to carry them yourself. With denser products weighing in excess of 100 lbs. (such as a 4 x 8 sheet), they can cause considerable strain on your body. The best advice for handling sheet goods is to never try to move full sheets by yourself. For those occasions when no help is available, there are a few tricks and devices you can employ to make carrying sheet goods less risky to the objects around them and to the general health of your muscles and back. Specially designed carts, like a rolling sheet goods cart (see photo, right) and a sheet goods dolly (see photo, next page), allow you to move the panels easily around your shop (although loading the panels into the devices can be tricky). There are also a number of handling aids, mostly shop-built, that can function like a third hand to help improve your grip and balance when carrying (one of these is shown in the photo at the bottom of the next page).

Storing sheet goods requires some care and planning. If left unchecked, it doesn't take many sheets of plywood to overwhelm your shop or to cut off access to any other building materials stored behind them. If you have the floor space and will be storing the panels for more than a few days before using them, laying them flat on a hard surface is the best method. This reduces the risk of warping that can occur when panels are stacked on edge. If you must store panels on edge, avoid leaning other objects against them, and stack them so they are as close to vertical as you can get them without risk of falling over. Slip scrap wood beneath the bottom edges to protect them from ground contact.

A rolling sheet goods storage cart is one of the handiest additions you can build for your workshop (see project plans, pages 197–203). The cart allows you store many panels in one convenient spot, where damage is less likely. And because it rolls, you can use it to transport the sheet goods to your work area. Shelving, bins, and an upper shelf create storage for dimension lumber and larger cutoff pieces.

How Much Do Sheet Goods Weigh?

Type	Size	Weight
Veneer core plywood (softwood face)	¾" x 4' x 8'	65 to 70 lbs.
Veneer core plywood (hardwood face)	¾" x 4' x 8'	55 to 60 lbs.
MDF	¾" x 4' x 8'	105 to 110 lbs.
Particleboard	¾" x 4' x 8'	90 to 100 lbs.
Composite-core plywood	¾" x 4' x 8'	80 to 85 lbs.
Lumber-core plywood	¾" x 4' x 8'	55 to 65 lbs.

Sheet Goods Tote Lends a Hand When Carrying Panels by Yourself

Although some types of sheet goods can get quite heavy (over 100 lbs. for a 4 x 8 sheet), the real problem with carrying them by yourself is that they are awkward. Unless you're a retired pro basketball player, you probably have a difficult time spanning the distance from edge to edge with your arms, and you won't find many handles on a sheet of MDF. The shop-built tote shown here is one of many handling aids that can help you get a better grip on bulky sheet goods. To make this tote, you'll need a 12"-long piece of 2 x 2 and two pieces of ¾"-thick plywood (one 2¾" x 12", the other 12" x 24"). Sandwich the 2 x 2 between the plywood pieces as shown, and join them with glue and 1½" wood or wallboard screws. Then make a handle by face-gluing two pieces of ¾" x 3" x 9" plywood, then cutting out the handle shape with a jigsaw. Center the ends of the handle on the back of the larger piece of plywood, 3" down from the top, and attach with glue and four 1½" screws driven though the plywood and into each handle end.

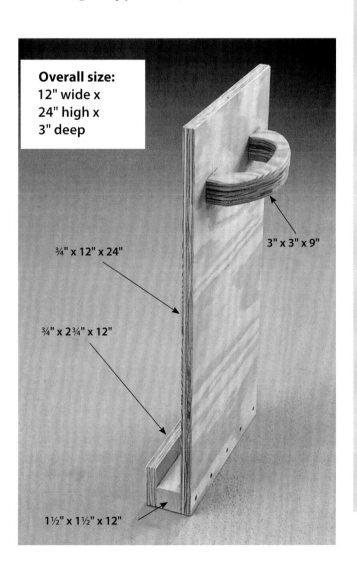

Overall size:
12" wide x
24" high x
3" deep

¾" x 12" x 24"

3" x 3" x 9"

¾" x 2¾" x 12"

1½" x 1½" x 12"

Sheet Goods Dolly

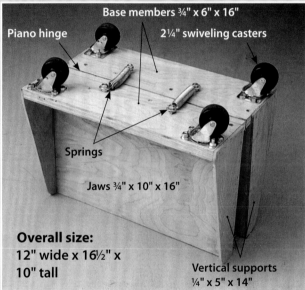

Base members ¾" x 6" x 16"

Piano hinge

2¼" swiveling casters

Springs

Jaws ¾" x 10" x 16"

Overall size:
12" wide x 16½" x
10" tall

Vertical supports
¼" x 5" x 14"

This compact sheet goods dolly built from scrap plywood actually uses the weight of the sheet goods themselves to pin them between the jaws of the dolly. The two halves of the dolly are held together with a piano hinge that flexes downward from the panel weight, drawing the jaws together. When the panels are removed, springs mounted on the undersides of the base members open the jaws up again for easy loading access of the next sheet goods "passenger."

DESIGN BY MICHAEL GUIMONT, FROM PRINCETON, MINNESOTA.

JOINERY

Joining sheet goods panels is a slightly different exercise than joining solid lumber. Many types of panels cannot be effectively machined to create such common woodworking joints as mortise-and-tenons, finger joints, or dovetail joints. Neither are they particularly capable when it comes to holding screws and nails (although some are certainly better than others). When using veneer-core plywood, for example, you can often get away with making glued butt joints that are reinforced with wood screws (see photo, right). MDF and particleboard generally hold nails better than screws, but, as with plywood, the glue will be doing most of the holding. You can increase glue surface by using dado, rabbet, or dado/rabbet joints (see below). You'll have more luck machining dadoes and rabbets into better-quality sheet goods.

Due to the ineffectiveness of typical woodworking joinery, mechanical fasteners often are used when building projects with sheet goods. Known generically as *knockdown (KD)*, *ready-to-assemble (RTA)*, or *European* hardware or fittings, mechanical fasteners are two-part pieces of hardware that rely on pressure on the opposing parts to make a joint. These fasteners have the added advantage of allowing your project to be easily disassembled and reassembled.

The basic butt joint is the workhorse of sheet goods joinery. Butt joints should be reinforced with wood glue and fasteners. Here, countersunk wood screws are driven to reinforce the joints between two plywood panels. You can also use biscuits, dowels, pocket screws, or mechanical knockdown (also called *ready-to-assemble*) fasteners. Mechanical fasteners generally are not used with glue.

Woodworking Joints Commonly Used to Join Sheet Goods

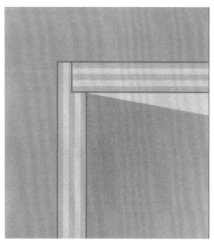

Dado joints are used mostly to join fixed shelves to cabinet sides or shelf standards. When using ¾"-thick panels, the dado should be ¾" wide x ⅜" deep—never cut a dado more than halfway into your stock. Dadoes are usually cut with a router and straight bit, or on a table saw fitted with a dado-blade set.

Rabbet joints are cut into the end or side of a panel to house the top or bottom panel of a carcase. Rabbets also are cut around the inside of a back opening to create a recess for the back panel. A router with a piloted rabbet bit is the best tool for cutting rabbets. You can also pre-cut rabbets into the individual panels before assembling the carcase.

Dado/rabbet joints create an interlocking joint at the top or bottom of a cabinet carcase. The tongue created when you rabbet one panel fits into a ⅜" x ⅜" dado cut into the other panel. In addition to the extra strength created by the interlocking nature of the joint, the dado/rabbet also strengthens a joint by creating additional surface for gluing.

Common Mechanical Fasteners (Shown in Cross-Section)

Minifix brand fittings are two-part fastening systems very similar to those used on mass-produced knockdown furnishings. The cam fits into a 15-mm hole drilled into the horizontal member and the screw assembly is fitted into an 8-mm hole in the vertical member. The screw head is inserted into the cam, and a setscrew in the cam is tightened to twist the cam and draw the parts together. A decorative cap is then snapped over the cam opening.

Screw assembly

Cam

Blum brand barrel-type two-part fasteners are used mainly to attach shelves in cabinets. The threaded nylon barrel is fitted into a 25-mm hole in the cabinet side, then the collared screw is driven into the end of the shelf so the duple-type head projects out. The head fits into the barrel, which is tightened by turning a screw-activated metal gripping plate.

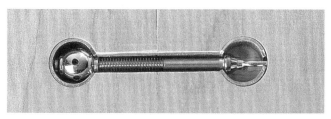

Tite-joint brand fasteners are especially useful for joining two sheet goods panels end to end, as with countertops. The heads of the fastener are mortised into adjoining panels. The mortises are connected by a groove or a guide hole that houses a threaded shaft. The sphere (the left head above) contains grip-holes so it can be spun with a scratch awl or small Allen wrench to tighten the joint.

Cross-dowel fasteners can be used to reinforce right-angle corners and to support shelving. The threaded steel dowel is screwed into a barrel cam that's mortised into the horizontal member. With cross-dowels, joints can be fastened and unfastened repeatedly without stripping the screw guide hole.

How to Use a Pocket-Screw Jig to Reinforce Joints

Depth stop collar

1

Clamp the workpiece that will contain the screw starter holes (usually the rails on a face frame, as shown above) into the pocket-screw jig. The center of the jig should align with the centerline of the workpiece. Mount the step drill bit that came with the jig into a portable drill, then drill through the guide bushings and into the workpiece until the depth stop makes contact with the mouth of the guide.

2

Clamp the workpiece containing the starter holes to the mating workpiece, then drive pocket screws through the starter holes and into the mating workpiece. Pocket screws are sold in packets at woodworking stores. They're thinner than regular wood screws to prevent the workpieces from splitting when they're driven. Most have square drive heads. Take care not to overdrive the screws.

Iron-On Edge Banding

Veneer edge banding is applied to conceal the exposed edges of sheet goods panels so they more closely resemble solid lumber. You can purchase veneer edge banding to match the species of most common plywood face veneers. Throughout this book, we use iron-on edge banding that has heat-activated glue pre-applied to the back of the tape. You can also purchase edge banding without pre-applied adhesives, this type is affixed with contact cement applied to the tape and to the edge of the workpiece.

For ¾"-thick plywood, use ¹³⁄₁₆"-wide tape (the extra width allows you to trim the tape to fit exactly). When taping shelving, apply the tape before installing the shelving.

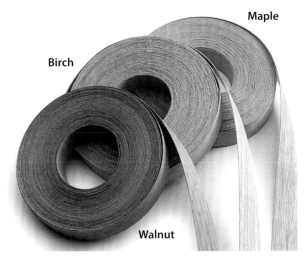

Maple

Birch

Walnut

Iron-on edge banding is sold in common plywood face veneer species. It can be purchased in 8' lengths or in rolls of 50' or 250'.

How to Apply Iron-on Edge Banding

Begin by cutting the edging for vertical surfaces to length with a utility knife. Cut the pieces slightly overlong.

Cover the foot of a household iron with foil to protect the iron from the glue. Set the iron to a low heat setting. With the veneer strip in position, press with the iron, moving away from one corner.

To guarantee a good glue bond, roll the edge banding with a wallpaper seam roller after each piece is applied. Try to work the roller in one direction only in case there are any trapped air bubbles.

Once the vertical pieces are applied, measure the horizontal pieces for the top, bottom, and fixed shelves. Cut these pieces to exact length using a square to guide the cut.

When all the pieces are applied, trim the overhang. Work in the same direction as the grain. We used a special edge-banding trimmer, but you can also use a cabinet scraper or a sharp utility knife.

SHEET VENEER

Choosing and applying your own wood veneer to your sheet goods project creates a host of new options in exotic wood species, distinctive grain figure, and in-grain pattern matching. Veneer can be applied to a flat substrate or to surfaces that are bent or shaped during the project construction.

The following photos illustrate how to apply non-adhesive-backed wood veneer. For many projects, you may prefer to use adhesive-backed veneer designed mostly for refacing cabinets. To apply adhesive-backed veneer, match the sheets as discussed in step 1, below, then simply remove the backing, position the veneer, and roll with a J-roller or wallpaper roller. Trim off the excess as in step 4, below.

Exotic and distinctive veneer types include: (A) zebrawood; (B) birdseye maple; (C) African padouk (vermillion); (D) madrone burl; (E) maple burl; (F) purpleheart. All shown with oil finish.

How to Apply Sheet Veneer to a Panel

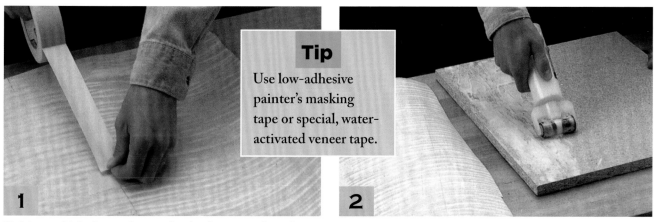

> ### Tip
> Use low-adhesive painter's masking tape or special, water-activated veneer tape.

1 If joining multiple sheets of veneer, lay the sheets together edge to edge in your chosen pattern. Use short strips of tape to hold the joint tight. Then lay a strip of tape down the entire length of the seam.

2 Apply an even, light coat of glue to the face of the substrate (we used a glue roller). Because the glue sets up quickly, have the veneer and your veneer press close at hand. If you're working with veneer for the first time, you may want to use white glue, which has a longer open working time.

3 Clamp the veneer to the glued-up panel in a veneer press—we made one from clamps, particleboard "plates," and wood strips (called cauls). Allow a few minutes for the glue to squeeze out, then check to see if the clamps need tightening. Let the glue dry for at least four hours before you remove the veneered panel from the press.

4 Trim off the excess veneer with a router or a laminate trimmer and a piloted flush-cutting bit. You can also use a sharp utility knife if you work very carefully.

HOW TO SQUARE UP A CARCASE

1

Test-fit the carcase side, front, and back panels by fitting and clamping them together. Adjust the parts as needed to achieve square corners and flush joints.

2

Apply wood glue to the edges of the parts, then clamp the parts together to form the carcase. Wood cauls (straight strips of scrap wood) distribute the clamp pressure evenly.

3

Test the carcase to see if it's square, using a framing square or by measuring diagonally from opposing corners—when the diagonal measurements are equal, the carcase is square, and you can go ahead and reinforce the joints with wood screws.

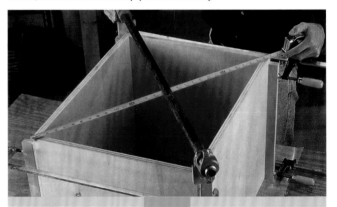

Tip

Adjust the carcase by applying a bar clamp or pipe clamp along one of the diagonals. Tightening the clamp will cause the carcase to "rack" slightly in the direction of pressure. You can also push or pull on the clamp heads to make adjustments.

Making Face Frames

Traditional cabinetry projects are designed with a hardwood face frame that's mounted to the front opening of the cabinet. The face frame conceals the edges of the carcase, provides bearing surface for hinges, and generally dresses up the cabinet (many contemporary cabinet styles do not employ face frames). Generally, the rails (horizontal members) and stiles (vertical members) of a face frame are butted together, not mitered. The stiles are allowed to run the entire height of the cabinet, with the ends of the rails butted against them. Glue and dowels, pocket screws, and splines are normally used to reinforce the joints.

Face frames are built independently from the cabinet carcase so they can be squared up accurately before they're mounted.

APPLYING PLASTIC LAMINATE

Durable, inexpensive, and available in many dozens of colors and styles, plastic laminate is a building product you should get to know. Used most frequently as a countertop or tabletop surface, it also can be applied to drawer fronts, cabinet doors, or even inside cabinets and drawers where a moisture-resistant, easy-to-clean surface is desirable.

Plastic laminate is sold in standard widths of 24", 36", 48", and 60" at building centers (or, simply look in your telephone book under the Counter Tops listing). You can order it in other sizes or even have it laminated onto a custom-sized substrate.

Particleboard and MDF are the two most common substrates for applying plastic laminate. Both have smooth surfaces that accept cement well. They are also very stable. But if your project will be exposed to constant moisture, use sanded, exterior-grade plywood for the substrate.

Tools and materials for working with plastic laminate include: (A) contact cement to bond laminate to substrate (use nonflammable product if working indoors or in an enclosed area); (B) J-roller; (C) paint roller with short-nap adhesive sleeve; (D) piloted flush-cutting router bit; (E) sample chips for making product selection; (F) disposable paintbrush for applying cement in tight areas.

How to Apply Plastic Laminate

1

The laminate sheet should overhang the edges of the substrate by ½" to 1" on all sides. Working on a flat, smooth surface, use a paint roller with a short-nap adhesive sleeve to roll a thin, even coat of contact cement onto the top surface of the substrate and the back face of the laminate.

2

To prevent the two cemented surfaces from bonding together while you position the laminate sheet, insert thin wood spacers at 6" intervals between the laminate and the substrate. Once the laminate sheet is positioned correctly, remove the spacers one at a time, starting at one end and working in order to the other end.

3

Use a J-roller to roll the laminate. Rolling creates a strong bond between the laminate and the substrate. Start in the middle of the workpiece and work toward the edges, rolling the J-roller in one direction only (this allows any trapped air bubbles to escape).

4

Trim off the excess laminate so the edge is flush with the substrate. For best results, use a router or laminate trimmer with a piloted flush-cutting bit. Smooth out any roughness from the router using a fine file.

MAKING AND INSTALLING SHELVES

Many woodworking projects built from sheet goods contain shelving. Because the edges of most sheet goods products are not meant to be left exposed, some type of shelf edge treatment is desirable. There are a number of ways to neatly and simply conceal the edges of shelving (see photo, right). When choosing which method to use, consider the complexity, the cost, and how well each type will meet your design standards. Make this decision up front, since it can have an effect on the dimensions of your project (if you are using solid shelf edge, for example, you'll normally want to reduce the width of your shelving by ¾").

The other shelving decision you'll need to make is choosing a method for supporting the shelves, particularly any adjustable shelves you may be building into the project design. A common practice when designing larger casework, like bookcases and entertainment centers, is to glue a fixed shelf (usually a middle shelf) into dadoes in the cabinet side, then install adjustable shelves above and below. The adjustable shelves typically employ pins or another type of shelf hardware. When designing your project, calculate your likely adjustable shelf heights so you can keep guide holes to a minimum.

(A) Solid-wood shelf edge can be purchased ready made in a variety of sizes, wood species, and profiles. Or, you can cut your own by shaping the edge of a board and ripping it to width on your table saw (see sequence, below).

(B) Iron-on edge tape made from matching wood veneer is a convenient, economical, and attractive product for treating shelf edges (see page 186).

(C) Exposed edge. For some types of projects made with higher-grade plywood (such as apple ply or Baltic birch), leaving the shelf edge untreated can be an effective design feature.

(D) Filled and painted. If you're planning to paint your woodworking project, you can simply fill any voids or imperfections in the shelf edge with wood putty or even joint compound, then sand the edge smooth before painting.

(E) Plastic T-slot shelf edge trim can be used to cover the exposed edges of shelving. Available at most woodworking or cabinet supply stores, it fits into a T-slot in the edge (the slot can be cut with a router or on your table saw). T-slot trim has definite commercial characteristics and the color selection is usually limited to white, black, or brown.

How to Make Custom Shelf Edge

1. Choose a board that's the same thickness as the shelving material and the same species as the face veneer. Mount an edge-forming bit into your router table (a single or double Roman ogee bit is a good choice) and shape the edge of the board.

2. Rip-cut the board to width on your table saw. If your shelving is ¾" thick, rip the shelf edge so it's ¾" x ¾". To make more shelf edge, shape the freshly cut edge, then rip to width again.

Options for Supporting Shelves

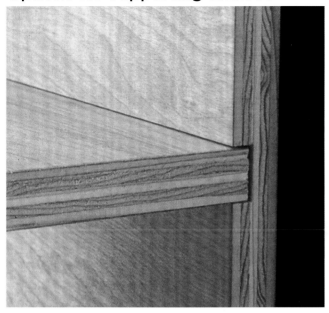

Dado grooves cut to the thickness of the shelves (¾" wide, typically) can be cut up to halfway through the thickness of the cabinet side or shelf standard (⅜" for ¾" stock). The dado provides a very sturdy bed for the shelf, especially when reinforced with glue and finish nails. But if the fit is too tight or the wood moves, bowing or breaking of the joint can occur. Often, a dado is used for the fixed shelf or shelves in a project only.

Shelf brackets recessed into grooves in the cabinet side or standard can be connected mechanically to shelf support tabs. This type of system provides strong support and plenty of shelf positioning options. The main drawback is their appearance: in most cases, the brackets will be visible when viewing your project from the front. Cutting the grooves also takes a little patience, and getting the slots in the brackets to line up can be tricky.

Shelf shown in cutaway

Shelf shown in cutaway

Shelf shown in cutaway

Shelf pins are made in many sizes, styles, and materials. The brass pins with mating grommets shown above are on the higher end of the shelf-pin spectrum. The grommet prevents the weight of the shelf from causing the support pin to ream out the guide hole. Use a piece of perforated hardboard as a drilling guide for locating guide holes.

Dowel pins are very economical to use. The photo above shows fluted, ¼"-diameter, dowel pins. You can make your own pins simply by cutting doweling to length (be sure to use hardwood doweling, however). If you rest the shelf directly on the dowel pins, it can roll, so cut dowel recesses in the shelf ends with a router and straight bit.

Plastic clips are inexpensive and reasonably sturdy. They're inserted into guide holes like those drilled for shelf pins, but the shape of the clips transmits part of the shelf load onto the cabinet sides.

HANGING DRAWERS AND DOORS

Mounting drawers and cabinet doors is one of the last steps in a woodworking project.

When it comes to hanging drawers, you can spend a lot of time building custom wood slides and glides, or you can purchase metal drawer slides that are sized to match the drawers and drawer openings in your project. By the same token, you can use a combination of traditional butt hinges and latches to hang doors on your cabinetry project; or you may prefer to try some of the contemporary and European self-closing hinges that most cabinetmakers have come to depend on. These newer products are usually easier to install and almost always easier to adjust, eliminating most of the headaches associated with hanging cabinet doors and drawers.

As a rule, decide which kind of hinges, slides, and hangers you'll be using before you finalize your project design.

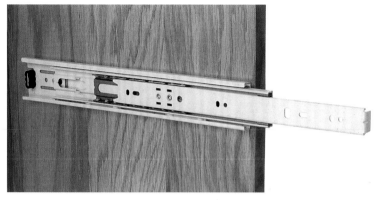

Side-mounted drawer glides are installed in cabinets and casework that do not have a face frame.

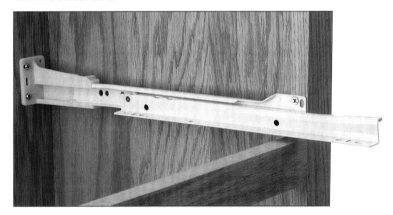

Rear/front-mounted drawer glides are installed in cabinets and casework that do have a face frame. The rear mounting bracket is sold as an accessory for most side-mounted glides.

How to Hang a Drawer Using Metal Slides

1

Mount the inner half of each drawer glide assembly to a drawer side. Follow the manufacturer's instructions for spacing. With some hardware, you may need to trim the back end to fit.

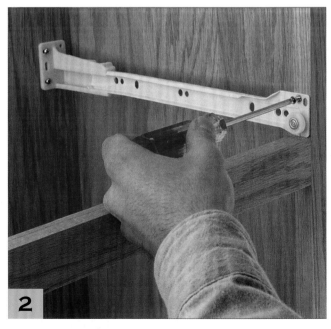

2

Mount the outer half of the glide mechanism to the inside of the cabinet, according to the manufacturer's directions for spacing. Most drawer glides have adjustable screw holes so you can locate the glide precisely where you want, after testing the fit by inserting the drawer into the drawer opening.

Cabinet Door Types

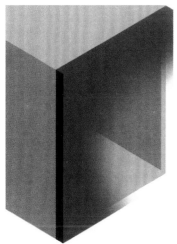

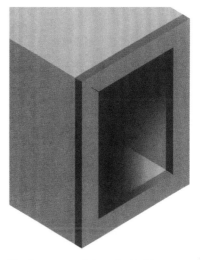

Full overlay door is flush with outer edges of face frame (or cabinet sides if no face frame is used).

Half overlay door closes against the outer face of the face frame, but does not fully obscure the face frame when closed.

Flush-mounted door fits inside the face frame opening for a more contemporary appearance.

How to Hang a Wall-Hung Cabinet with French Cleats

3" wood screws driven at wall stud locations

French cleats are made by bevel-ripping a board that's at least 4" wide, then fastening one half of the board to the wall and the other to the back of your cabinet. The cleats distribute the load of the cabinet over a broad area, reducing the strain on individual screws and the cabinet structure.

Bevel-rip a piece of ¾" plywood or 1" (nominal) dimension lumber at 45°. Attach one half to the back of your cabinet, with the bevel facing down. The back panel of the cabinet should be recessed into the sides to allow for the thickness of the cleat. Attach the other half of the cleat to the wall, using two 3" wood screws driven at each wall stud location spanned by the cleat.

Test the wall-mounted half of the cleat to make sure it's securely fastened to the wall, then lift the cabinet over the cleat and slide it down so the cabinet-mounted half fits over the wall-mounted half of the cleat. Check with a level and adjust, then drive a couple of screws through the back panel and into the wall to keep the cabinet from shifting or falling if it is bumped.

Applying a Finish

A lot of people are intimidated by the prospect of applying a finish to their carefully crafted woodworking project. Whether you're dealing with a plywood armoire or a solid walnut bandsaw box, the keys to getting a satisfactory finish are the same: do careful, thorough prep work; make a wise product choice; and follow the product manufacturer's directions closely when applying the product. When clear-finishing or staining a plywood project, condition the wood first (see photos, below).

Fill nail and screw holes, voids in plywood edges, and other surface defects with paintable or stainable wood putty. Overfill the area slightly, then sand the putty so it's even with the surface of the adjoining wood once the putty has dried.

Tips for Sanding Sheet Goods

Use a random-orbit sander with dust extraction for most of your finish-sanding. This type of sander leaves minimal sanding marks. With sheet goods, it's seldom necessary to sand past 150 grit.

Avoid over-sanding. Face veneer on most American and Canadian produced plywood is about 1/32" thick. Even when using medium or fine sandpaper, it doesn't take long to sand through the face veneer.

Wood Surface Preparations Methods Compared

Liquid stain applied over untreated pine veneer looks blotchy and dark and is hard to control.

Liquid stain applied over pine treated with a wash coat of commercial wood conditioner (can use diluted shellac instead) has even color penetration and is lighter in tone.

Gel stain applied over untreated pine also provides even color penetration since gel stains do not penetrate wood surfaces as deeply as liquid stains.

Visual Reference Chart: Common Wood Stain Tones Applied to Plywood

Pine Plywood

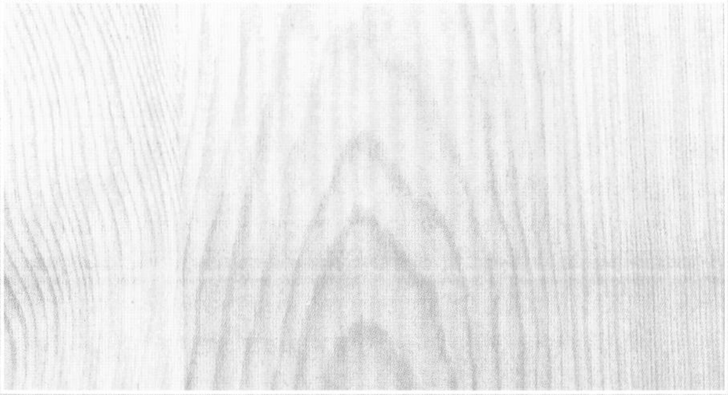

Clear topcoat only

Light stain

Medium stain

Dark stain

Oak Plywood

Clear topcoat only

Light stain

Medium stain

Dark stain

PART 3
Projects

SHEET GOODS CART

Every shop can benefit by increasing storage and reducing clutter. If wall space is at a premium in your shop, our sheet goods cart is a clever alternative to permanent shelving. Made of inexpensive CDX plywood, the cart occupies only 12'-square of floor space, yet it provides ample storage for full-sized sheet stock, center shelves for longer boards, and five bins of various widths for shorter cutoffs and scrap. The top shelf is perfect for storing containers of hardware or smaller tools, and one end of the cart sports notched holders for pipe or bar clamps. Casters allow you to roll the cart right where it's needed or out of the way entirely.

Vital Statistics: Sheet Goods Cart

Type: Rolling storage cart

Overall Size: 53" H x 72" L x 24" W at base

Material: CDX plywood

Joinery: Dadoes, screwed butt joints

Construction Details:

- Cart side that stores full sheets angled back 5° to keep sheets from tipping over
- Shelves between tall sides secured with dadoes and screws
- Storage bins separated by angled dividers
- Cart bottom reinforced with blocking to provide solid base for mounting casters
- Built-in clamp rack constructed from scrap CDX

Finish: None

Building Time

Preparing Stock: 2 hours

Layout: 2–3 hours

Cutting Parts: 3–4 hours

Assembly: 2–3 hours

Total: 9–12 hours

Tools

- Table saw
- Circular saw
- Drill/driver
- Sliding power miter saw (optional)
- Router and ¾" straight bit
- Jigsaw
- Clamps
- Sockets

Shopping List

- ○ (4) ¾" x 4' x 8' CDX plywood
- ○ (16) ⅜" x 2" carriage bolts, nuts, and washers
- ○ (2) 4" straight casters
- ○ (2) 4" swiveling casters with brakes
- ○ #8 flathead wood screws (1¼", 1½", 2¼")
- ○ Wood glue

Exploded View

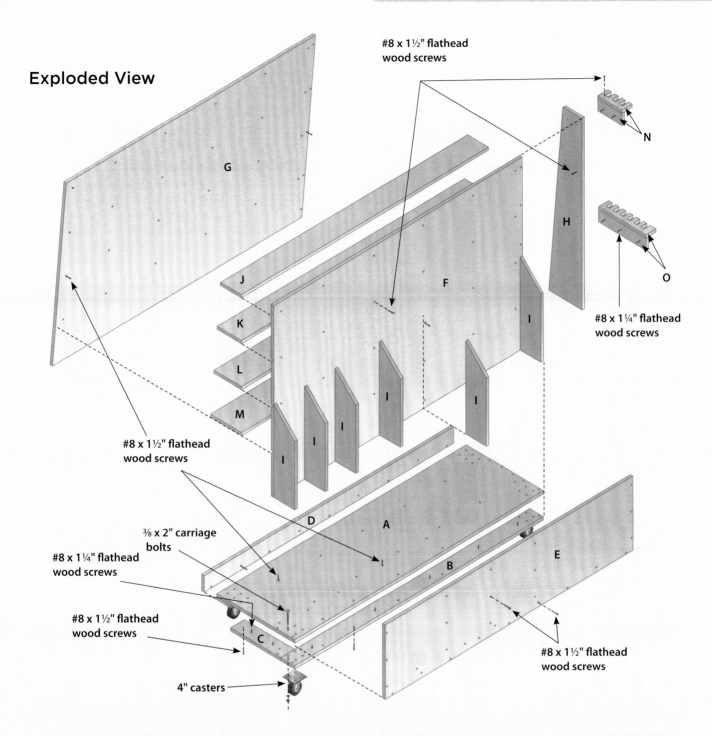

#8 x 1½" flathead wood screws

N

O

#8 x 1¼" flathead wood screws

G

J

K

L

M

H

F

I

I

I

I

I

I

#8 x 1½" flathead wood screws

D

A

B

E

⅜ x 2" carriage bolts

#8 x 1¼" flathead wood screws

#8 x 1½" flathead wood screws

C

4" casters

#8 x 1½" flathead wood screws

Sheet Goods Cart Cutting List

Part	No.	Size	Material	Part	No.	Size	Material
A. Base	1	¾" x 22½" x 72"	CDX plywood	**I.** Divider	6	¾" x 6" x 22½"	CDX plywood
B Blocking (long)	2	¾" x 4" x 72"	"	**J.** Top shelf	1	¾" x 6³⁄₁₆" x 71¼"	"
C. Blocking (short)	2	¾" x 4" x 14½"	"	**K.** Shelf	1	¾" x 7³⁄₁₆" x 71¼"	"
D. Short edge	1	¾" x 4" x 72"	"	**L.** Shelf	1	¾" x 8⅛" x 71¼"	"
E. Tall edge	1	¾" x 18" x 72"	"	**M.** Shelf	1	¾" x 9⅛" x 71¼"	"
F. Vertical side	1	¾" x 46½" x 72"	"	**N.** Clamp holder	2	¾" x 3" x 7"	"
G. Angled side	1	¾" x 46¾" x 72"	"	**O.** Clamp holder	2	¾" x 3" x 13¾"	"
H. End	1	¾" x 9¾" x 46½"	"				

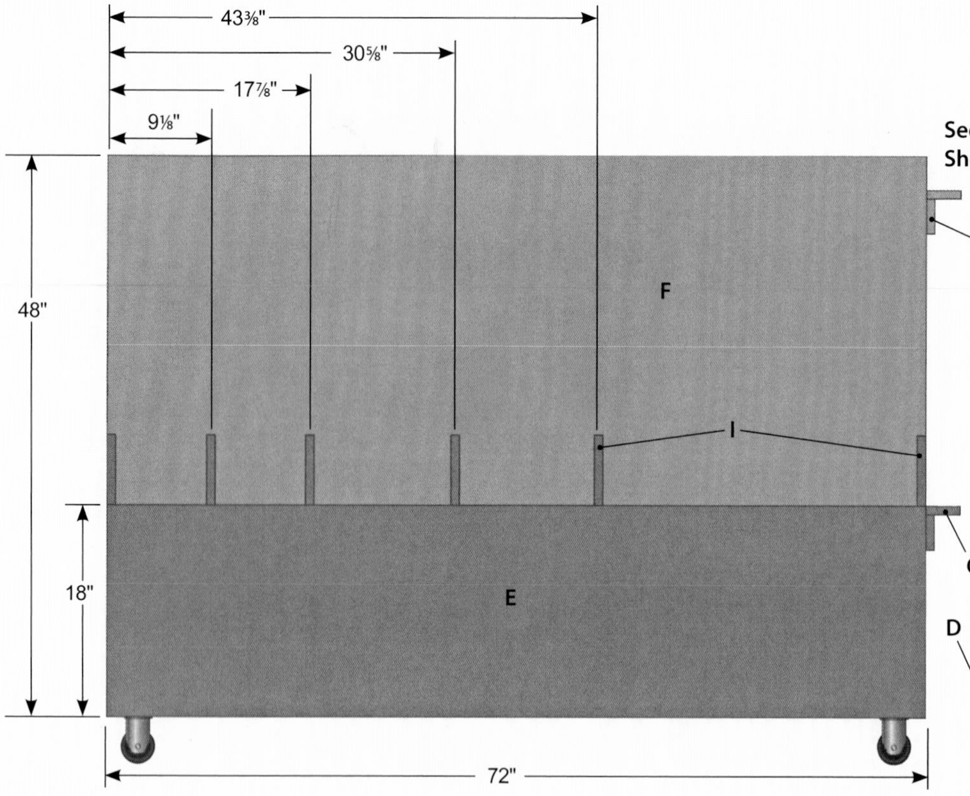
43⅜"
30⅝"
17⅞"
9⅛"
48"
18"
F
I
N
O
E
72"

FRONT VIEW (DIVIDED SIDE)

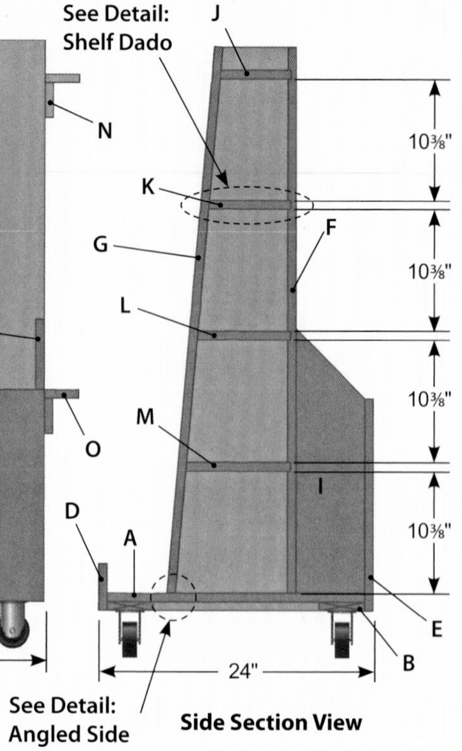
See Detail:
Shelf Dado
J
K
G
F
L
M
I
D
A
E
B
10⅜"
10⅜"
10⅜"
10⅜"
24"
See Detail:
Angled Side

Side Section View

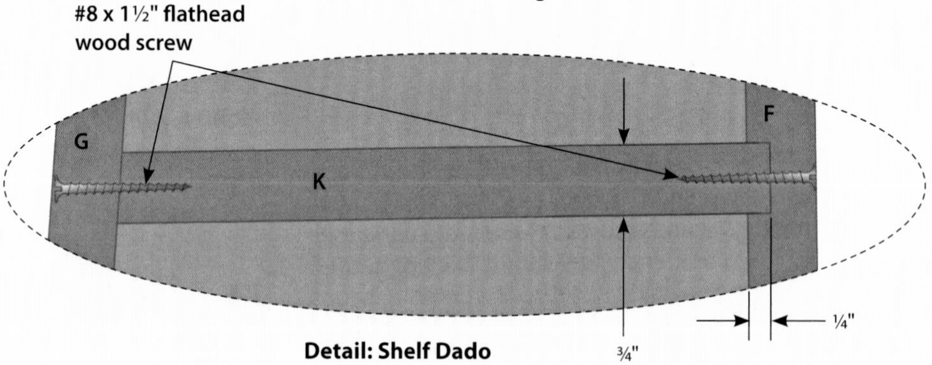
#8 x 1½" flathead
wood screw
G
F
K
¾"
¼"

Detail: Shelf Dado

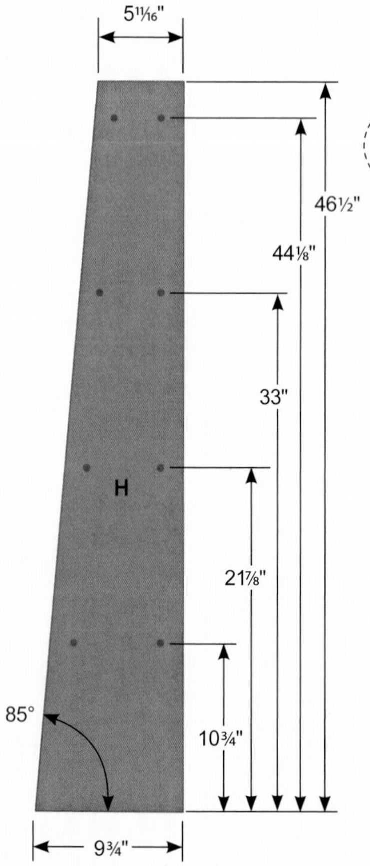
5¹¹⁄₁₆"
46½"
44⅛"
33"
21⅞"
10¾"
H
85°
9¾"

Drilling Pattern: End

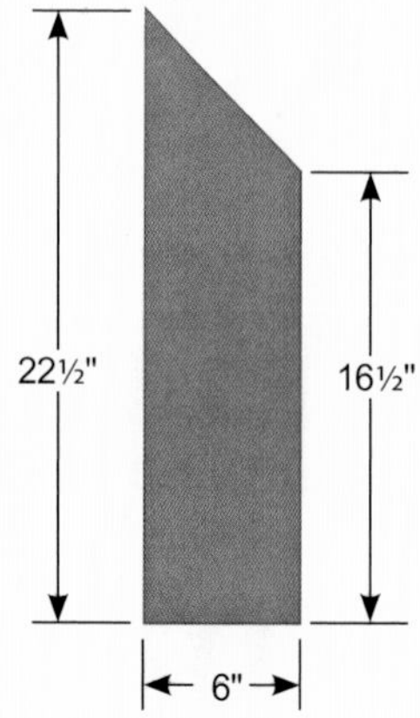
22½"
16½"
6"

Divider

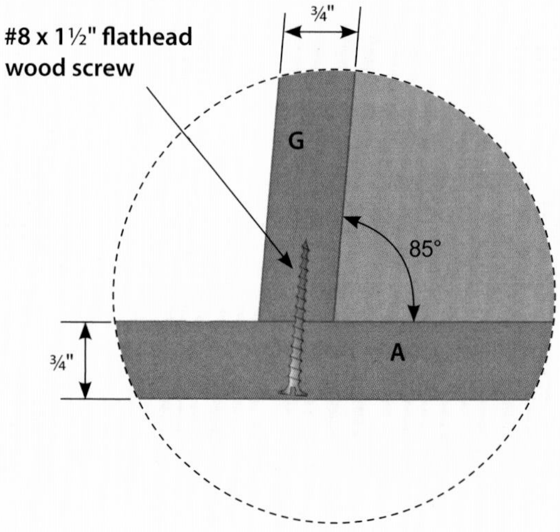
¾"
#8 x 1½" flathead
wood screw
G
85°
A
¾"

Detail: Angled Side

Sheet Goods Cart: Step-by-Step

Photo A: Attach the long and short blocking pieces to the bottom panel, flush around the perimeter, with glue and 1¼" flathead wood screws. Drill pilot holes for the screws first.

Build the Base

1 Cut to size the base, bottom blocking, and short and tall edge pieces from ¾" plywood. Use a circular saw and a straightedge guide for making the initial cuts to reduce full sheets to a more manageable size. (Remember to account for the offset between the blade and the saw foot as you line up the straightedge guide for your cuts.)

2 Lay the base on a flat worksurface and attach the short and long blocking to it. The ends of the short blocking pieces fit in between the longer blocking. Use glue and 1¼" flathead wood screws to fasten the parts (**see Photo A**). Mark two reference lines along the length of the base on the blocking side to serve as centerlines for attaching the tall side pieces later. Draw a line for the angled side 5⅝" from one edge, and draw another line for the vertical side 6⅜" from the other edge.

3 Attach the short and tall edges to the base assembly using glue and 1½" flathead wood screws. Flip the base assembly over before attaching the edge pieces so the blocking faces down. Align the ends of the parts and make sure the bottoms of the edge pieces are flush with the bottoms of the blocking. Drill pilot holes first, spacing the screws about every 8". Alternate the screws between the base and blocking to increase the joint strength. Mark centerlines on the outside of the tall edge for fastening the dividers. See Front View (Divided Side), page 200, for locating the dividers. Position the outermost divider lines ⅜" from the ends of the tall edge.

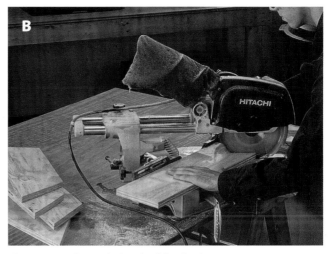

Photo B: Cut the angled ends of the dividers. A power miter saw makes this task quick and easy, once you've established the cutting angle. You can also make these cuts with a jigsaw, table saw, or circular saw.

Spacers

Photo C: Glue and screw the dividers to the tall edge with 1½" wood screws. Use scrap-wood spacers inserted between the dividers to establish divider spacing.

Install the Dividers

4 Cut the six dividers to size. Follow the measurements given on the Divider diagram, page 200, to mark the angled ends. We used a power miter saw to cut the dividers (**see Photo B**), but you could also use a circular saw, table saw, or handsaw to make these cuts.

5 Attach the dividers to the tall edge piece. The outer dividers are attached to the tall edge using glue and 1½" screws. Keep the outside faces of the outer dividers flush with the ends of the tall edge. Cut two scrap-plywood spacers, 8" and 12" wide, and insert a spacer between each pair of dividers as you attach the dividers with glue and screws (**see Photo C**). Use the centerlines you drew on the outside of the tall edge for lining up the screws. Then extend the divider centerlines down around the bottom blocking, and drive two 2¼" screws up through the cart base and into the bottom of each divider.

Photo D: Cut shelf dadoes in the vertical side with a router and straightedge guide. Pull the router toward you as you make each cut, being careful to hold the router base tight against the straightedge as you work.

Assemble the Center Section

6 Rip-cut and crosscut the angled side and the four shelves to size. For each of these parts, tilt the saw blade 5° to create a bevel along one long edge. Mark the beveled edges on the parts to keep the orientation clear later.

7 Cut the vertical side to size and rout the shelf dadoes into one face. The dadoes are ¾" wide and ¼" deep. Mark a set of long reference lines for each of the four dadoes using the Side Section View, page 200, to place the dadoes. Then extend dado centerlines to the other face of the vertical side to serve as screw guidelines. Cut the dadoes with a router and a ¾" straight bit (**see Photo D**). Clamp a straightedge on the vertical side to guide the router. To line up the guide, measure the distance from the edge of the router bit to the outer edge of the router base. This is the distance the straightedge must be offset from the closest marked dado line of each cut. Reset the straightedge for cutting each dado.

8 Attach the vertical side. Spread glue onto the edges of the dividers and clamp the vertical side in place so the dadoes face away from the dividers. Drive 1½" flathead wood screws through the vertical side into pilot holes in the two end dividers. Then, using your 8" and 12" spacers between the dividers as alignment aids, screw the vertical side to the inside dividers. Tip the cart assembly onto the face of the tall edge and drive 1½" screws along the vertical side reference line to attach the vertical side to the base.

9 Attach the shelves to the vertical side (**see Photo E**). Cut eight 10⅜"-long scrap spacers to support the ends of the shelves during assembly. Spread glue in the bottom dado and insert the square edge of the bottom shelf into the dado. With the divider side of the cart facing you, keep the end of the shelf flush

Photo E: Fasten the shelves to the vertical side with glue and screws. Support the shelves with scrap-wood spacers. Insert the square edge of the shelves into the dadoes, and keep the beveled edges facing up.

Photo F: Draw centerlines for the shelves on the angled side, and then fasten the angled side to the shelves with glue and screws. Screw up through the bottom to attach the angled side from below.

with the left end of the vertical side. The dado joint for this bottom shelf will be fastened with glue only. Install this shelf so the beveled edge faces up (see Detail: Shelf Dado, page 200) and clamp the shelf. Then install the rest of the shelves into the dadoes with 1½" wood screws and glue. Arrange each shelf so the beveled edge faces up, and support the shelves with spacers. Make sure the left shelf ends are flush with the end of the vertical side.

10 et the angled side against the shelves and clamp it in place temporarily. Draw shelf centerlines across the face of the angled side, then remove the shelf spacers. Glue and fasten the angled side to the shelves with 1½" screws (**see Photo F**). Screw the cart base to the angled side, following the angled side reference line you drew in step 2 under "Build the Base" on page 201.

Photo G: Attach the casters. Mark bolt hole locations on the corners of the cart bottom, positioning the casters so they sit squarely on the bottom blocking. Drill the holes and bolt the casters in place, with the washers and nuts facing the cart bottom.

11 Lay out and cut the end piece to size, using the measurements given in Drilling Pattern: End, page 200. Mark the screw locations on the end piece. Set the end piece into position on the end of the cart where the shelves are set back from the ends of the side panels. Attach the end piece to the ends of the shelves with glue and 1½" wood screws. Fasten the end piece with 2¼" screws driven up through the base and blocking.

Install the Casters

12 Tip the cart on its side and install the four casters. Lay the base of each caster in place on the blocking pieces and use the caster base holes to mark locations for carriage bolts. Position the casters so that all four corners of each caster rest firmly on the bottom blocking. Also, be sure the caster holes will not interfere with screws attaching the bottom blocking or dividers. Drill ⅜" pilot holes for each caster. Install the two straight casters on one end and swiveling casters on the other end, with the washers and nuts facing the caster wheels (**see Photo G**).

Finishing Touches

Expand the storage possibilities of your cart by adding clamp holders to one end. We made ours out of scrap CDX left over from the project. The holder configuration you choose will depend on the number, style, and length of clamps you have. We've designed the storage cart tall enough to hang several 4'-long pipe or bar clamps without the clamps dragging on the floor. The bottom holder has additional cutouts for shorter clamps.

13 Make the clamp holders. Each clamp holder is composed of two parts that form an "L" when the parts are assembled. Mark cutouts on the top member of each holder and cut them out with a jigsaw (**see Photo H**). For standard pipe

Photo H: Lay out the clamp holders and cut them to size. Notches 1" wide and 2" deep are a good size for holding standard pipe clamps. The bottom clamp holder can be made longer than the top, spanning across the end piece and the end divider.

Photo I: Fasten the clamp holders to scrap blocking, and screw the blocking to the end of the cart with 1½" wood screws. Keep the slots aligned between the holders so the clamps will hang straight.

clamps, 1"-wide cutouts 2" deep will hold each clamp securely. Assemble the holders with glue and 1½" screws. Then attach the holders to the closed end of the cart using glue and 1½" screws (**see Photo I**). Make sure to keep the top and bottom clamp holders aligned when fastening them to the cart so long clamps will hang straight.

14 Break all exposed sharp edges with sandpaper to minimize splinters. We chose not to apply any finish to our sheet goods cart, but you may prefer to dress yours up with a couple coats of enamel paint. If you plan to store veneered plywood sheets with finished faces on your cart, you may want to add strips of carpet to the face of the angled side to protect the veneer from scratches.

Jobsite Table Saw Mobile Base

While many jobsite table saws are equipped with folding stands, none offer a handy storage drawer like this project does for keeping a dado blade set, push sticks, push pads, and other saw accessories within easy reach. Its shallow top tray will capture residual dust that the saw's dust port may not, to help keep your saw and work area cleaner—just pull it out and vacuum up the sawdust when needed. Casters will make your saw and mobile base easier to roll around the shop, wherever it's needed. This mobile base will raise most portable table saws to a comfortable overall working height of around 36".

Vital Statistics: Jobsite Table Saw Mobile Base

Type: Rolling storage base

Overall Size: 25" W x 23" D x 23" T

Material: Baltic birch plywood

Joinery: Butt joints, rabbet joints

Construction Details:

- Sized to raise a jobsite table saw to 36" working height

- Top shallow pullout tray captures residual sawdust that saw's dust port does not

- Deep bottom drawer for saw accessory storage (push pads, dado blade, featherboards, etc.)

- Saw mounts to top framework with screws

Finish: Project can be left unfinished or given a protective topcoat for added durability

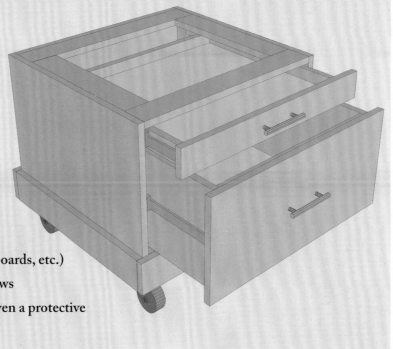

Building Time

 Preparing Stock: 1 hour

 Layout: 2 hours

 Cutting Parts: 3 hours

 Assembly: 2 hours

Total: 8 hours

Tools

- Table saw
- Drill/driver
- Pocket hold jig
- Clamps
- Rachet, sockets

Shopping List

- ❍ Baltic birch plywood (¼", ¾")
- ❍ (4) 20" full-extension drawer slides
- ❍ (4) 5" swiveling, locking casters
- ❍ (2) 5" bar pulls, nickel finish
- ❍ 1¼" brad nails
- ❍ Wood screws (1¼", 2")
- ❍ Wood glue
- ❍ 1¼" pocket screws
- ❍ ¼" x 1" lag screws and washers

Exploded View

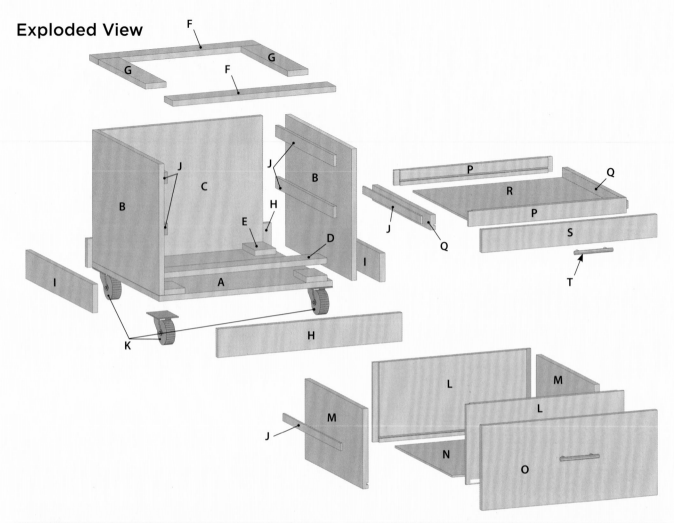

Main Base Cutting List

	Part	No.	Size	Material
A.	Bottom	1	¾" x 23" x 25"	Baltic birch plywood
B.	Sides	2	¾" x 17¼" x 23"	"
C.	Back	1	¾" x 17¼" x 23½"	"
D.	Brace	1	¾" x 3" x 23½"	"
E.	Caster blocking	4	¾" x 3" x 4½"	"
F.	Mounting plates, long	2	¾" x 3" x 23½"	"
G.	Mounting plates, short	2	¾" x 3" x 16¼"	"
H.	Skirt boards, front/back	2	¾" x 3¾" x 26½"	"
I.	Skirt boards, side	2	¾" x 3¾" x 23"	"
J.	Drawer slides	4	20"	Full-extension
K.	Casters	4	5"	Swiveling, locking

Drawers Cutting List

	Part	No.	Size	Material
L.	Drawer front, back	2	¾" x 10½" x 22½"	Baltic birch plywood
M.	Drawer sides	2	¾" x 10½" x 20"	"
N.	Drawer bottom	1	¼" x 19¾" x 21½"	"
O.	Drawer face	1	¾" x 11⅛" x 25"	"
P.	Tray front, back	2	¾" x 2" x 22½"	"
Q.	Tray sides	2	¾" x 2" x 20"	"
R.	Tray bottom	1	¼" x 19¾" x 21½"	"
S.	Tray face	1	¾" x 2" x 25"	"
T.	Bar pull handles	2	5"	"

Jobsite Table Saw Mobile Base: Step-by-Step

Photo A: Assemble the bottom, side, and back panels with glue and 2" countersunk screws.

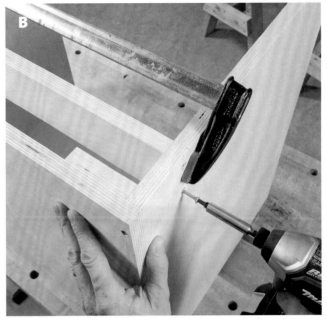

Photo B: The front edges of the brace and side panels should be flush.

Build the Base

1 Rip-cut and crosscut the bottom, side, and back panels to size. Assemble the side and back panels with glue and 2" countersunk wood screws driven through the side panels and into the ends of the back panel. Attach the bottom panel to the side and back panels with more glue and screws (**see Photo A**).

2 Cut four caster blocking pieces to shape. Spread glue onto their bottom faces, and position the blocking against the inside corners of the bottom panel. Tack the blocking in place with 1¼" brad nails to hold the pieces in place while the glue dries.

3 Cut the brace to size and install it between the side panels with countersunk screws so the front edges of the parts are flush (**see Photo B**). Locate the top face of the brace 3" up from the bottom edges of the side panels; it will serve as an attachment point for the front skirt board and help to reinforce it from behind.

4 Prepare two long and two short mounting plates; these will be attachment points for the table saw's base. Assemble the mounting plates into a framework with pairs of 1¼" pocket screws driven through the ends of the short mounting plates and into the edges of the long mounting plates. Work carefully to ensure that these joints remain flush (**see Photo C**).

Photo C: Remember to keep the joints flush while assembling the mounting plates.

5 Set the framework into the mobile base, flush with its top edges. Drive 2" screws through the side and back panels to secure the mounting plate framework in the base (**see Photo D**).

6 Rip-cut and crosscut four skirt boards that will cover the edges of the mobile base's bottom panel. Install the side skirt boards first with glue and brad nails; their bottom edges should be held flush with the bottom panel's bottom face. Fasten the front and back skirt boards the same way, nailing the front skirt board to the brace and bottom panel (**see Photo E**). The ends of the front and back skirt boards overlap the ends of the side skirt boards.

7 Draw a pair of parallel layout lines across the inside face of each side panel from front to back to mark centerlines for installing drawer slide hardware inside the base. Locate these layout lines $1\frac{7}{8}$" and $8\frac{5}{8}$" down from the top ends of the side panels.

8 Unclip the "cabinet" side components from the "drawer" side components of the slide hardware. Install the "cabinet" side components inside the base with short screws that are supplied with the slides. Position the slides so their front edges are flush with the base front. Make sure the slides are carefully centered on the layout lines you drew in the previous step (**see Photo F**).

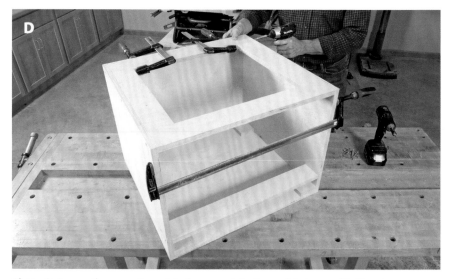

Photo D: Secure the mounting plate framework to the mobile base with 2" screws.

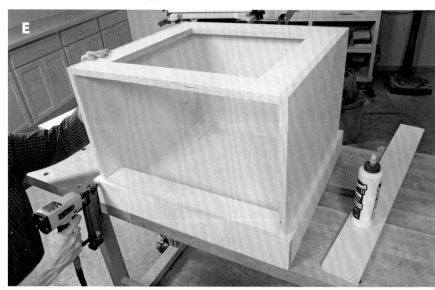

Photo E: Use glue, brad nails, and clamps while installing the four skirt boards.

Photo F: Install the slides with the short screws that come with the hardware, making sure the slides are centered on the parallel lines drawn in step 7.

Photo G: The casters need to be flush with the corners of the bottom panel.

Photo H: A miter gauge and a long scrap fence will help reduce tearout while cutting a ⅜"-deep, ¾"-wide rabbet into the drawer front and back panels.

Photo I: Use a rip fence ⅜" from the blade and a standard saw blade raised ¼" above the table to cut the first pass in the bottom panel's tray and drawer parts.

9 Turn the mobile base upside down so you can install the four swiveling casters on the bottom panel with ¼" x 1" lag screws and washers. Position the casters so their mounting plates are flush with the corners of the bottom panel (**see Photo G**).

Build the Drawers

10 Clip the drawer slide components together again. Measure the distance between the slides to confirm the final length of the front and back panels for the drawer and tray. Make a note of these lengths if they are different from the lengths provided in the Shopping List (see page 205).

11 Rip-cut and crosscut workpieces to size for the front, back, and side panels of the drawer and tray. Install a wide dado blade into your saw so you can cut a ⅜"-deep, ¾"-wide rabbet into the ends of the drawer front and back panels. Make test cuts first on a piece of scrap to ensure that the side panels will tuck neatly into the rabbets. With your saw's setting dialed in, cut the drawer and tray rabbets into the front and back panels. Support the workpieces from behind using your saw's miter gauge equipped with a long scrap fence to help minimize tearout from the dado blade (**see Photo H**).

12 Reinstall your saw's standard blade, and raise it ¼" above the table to prepare for cutting grooves in the tray and drawer parts for the bottom panels. Set the rip fence ⅜" from the blade, and make the first pass along the inside bottom face of the drawer and tray parts. Now reset the saw fence as needed to widen these grooves to fit the ¼" plywood you'll use for the bottom panels. Make a second pass alongside the first cut to bring the bottom panel grooves to their final width. Check that your plywood will fit the grooves you've made (**see Photo I**).

13 Cut two bottom panels of the same size for the drawer and tray. Dry-fit the fronts, backs, and sides of the drawer and tray with the bottom panels in their grooves to be sure the rabbet joints close properly. If they do, assemble the drawer and tray with glue and clamps. Measure across the diagonals of each of these assemblies before the glue sets to check that the drawer and tray are square. Adjust the clamps to correct for any deviation (**see Photo J**).

14 Reinforce the drawer and tray rabbet joints by driving several brad nails or countersunk wood screws through the tongues of the rabbets and into the side workpieces.

15 Draw centered layout lines along the sides of the drawer and tray from front to back so you can install the "drawer" side slide members. Install this hardware with the front edges of the components flush with the front of the drawer and tray. Engage the slide hardware to hang the drawer and tray inside the base. Check the action of the slides to make sure the drawer and tray open and close smoothly (**see Photo K**).

16 Rip-cut and crosscut face panels for the drawer and tray, then soften their corners and edges with a file or sanding block. Apply several strips of double-sided carpet tape to the front of the drawer and tray, then carefully set the face workpieces into place. Aim for an even gap of about 1⁄16" between the drawer and tray faces. Press the panels firmly into place. Carefully open both compartments so you can clamp the faces securely in place. Drive several 1¼" countersunk wood screws through the drawer and tray fronts into the faces to attach them permanently (**see Photo L**).

Photo J: Before the glue sets, a final measure across the diagonals of each drawer and tray assembly will ensure exactness.

Photo K: Always test that the drawers slide in and out smoothly.

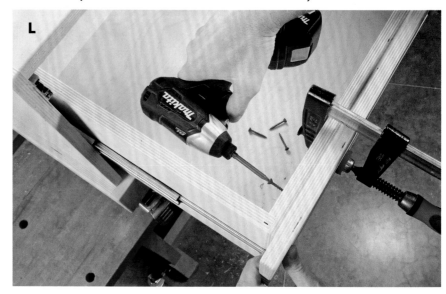

Photo L: Use 1¼" countersunk wood screws to permanently attach the faces to the drawers and tray fronts.

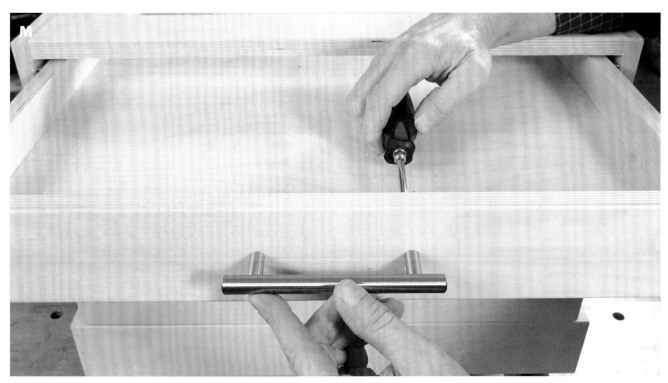

Photo M: Use long screws to attach the drawer pulls.

Photo N: The table saw should be installed with nuts through the mounting plate framework.

Finishing Touches

17 Mark the faces of the drawer and tray for a centered drawer pull. Bore a pair of holes through the face and front panels, and attach the pulls with long screws (**see Photo M**).

18 If you wish to add a protective finish, such as polyurethane, to your new mobile base, remove the slide and pull hardware. Sand all the project surfaces smooth and apply two coats of finish. When the finish dries thoroughly, install your table saw on the mobile base with bolts driven through the mounting plate framework and secured with nuts (**see Photo N**).

Outfeed Cart

The primary purpose for a cart like this is to support long or large workpieces when making rip cuts or sizing down sheets of plywood. What makes this project particularly handy is that the middle shelf and lower compartments can help you organize and store boards up to 5' or 6' long as well as shorter offcuts. Notice that the top of the cart overhangs the project by 12" along its back side; this way, you can roll the cart up behind a table saw with a rear dust collection port without interference, then connect a 4" dust collection hose to it and extend the hose between the casters on the floor.

Vital Statistics: Outfeed Cart

Type: Rolling storage cart

Overall Size: 53" H x 72" L x 24" W at base

Material: Maple or birch plywood, poplar lumber

Joinery: Screwed butt joints

Construction Details:

- Top overhangs rear of cart by 12" to clear cabinet saw dust collection port and 4" hose at bottom

- Middle shelf stores longer offcuts up to 60" long; bottom divided compartment stores shorter offcuts, tool bags or cases, etc.

Finish: Project can be left unfinished or given a protective topcoat for added durability

Building Time

Preparing Stock: 3 hours

Layout: 2 hours

Cutting Parts: 2 hours

Assembly: 3 hours

Total: 10 hours

Tools

- Table saw
- Drill/driver
- Clamp
- File, sanding block, or sander
- Handheld router
- ⅛" or ¼" radius roundover bit
- Sliding power miter saw (optional)
- Ratchet and sockets
- Jigsaw or band saw
- Clamps

Shopping List

- ○ ¾" plywood
- ○ 5" casters
- ○ (4) leg strips
- ○ Wood screws (1¼", 2")
- ○ Wood putty or wood plugs
- ○ Brad nails (1¼", 2")
- ○ ¼" x 1" lag screws and washers
- ○ Wood glue

Exploded View

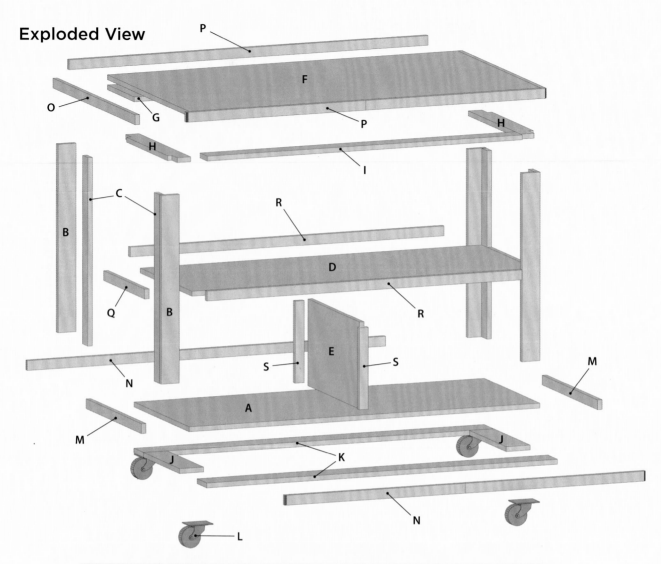

Outfeed Cart Cutting List

Part	No.	Size	Material
A. Bottom	1	¾" x 23¼" x 58½"	Plywood
B. Leg strips, wide	4	¾" x 3" x 28¾"	"
C. Leg strips, narrow	4	¾" x 2¼" x 28¾"	"
D. Shelf	1	¾" x 21¾" x 57"	"
E. Divider	1	¾" x 13¼" x 21¾"	"
F. Top	1	¾" x 34½" x 58½"	"
G. Overhang lamination	1	¾" x 34½" x 58½"	"
H. Build-up strips, side	2	¾" x 3" x 21¾"	"
I. Build-up strip, front	1	¾" x 3" x 52½"	"
J. Blocking, side	2	¾" x 4" x 15¼"	"
K. Blocking, front/back	1	¾" x 4" x 58½"	"
L. Casters	4	5"	Swiveling, fixed
M. Skirt boards, side	2	¾" x 1½" x 23¼"	Plywood
N. Skirt boards, front/back	2	¾" x 1½" x 60"	"
O. Top trim, side	2	¾" x 1½" x 34½"	"
P. Top trim, front/back	2	¾" x 1½" x 60"	"
Q. Shelf trim, side	2	¾" x 1½" x 17¼"	"
R. Shelf trim, front/back	2	¾" x 1½" x 52½"	"
S. Divider trim	2	¾" x 1½" x 12½"	"

Outfeed Cart: Step-by-Step

Construct the Base Framework

1 Rip-cut and crosscut the bottom panel to size (**see Photo A**).

2 Each of the cart's legs consists of a wide and narrow strip that form a 90º corner when assembled. Rip-cut and crosscut the eight leg strips, and arrange each pair of leg components so the edge of the narrow strip butts against a face of the wide strip, flush with its edge. Assemble the four legs with glue and 2" screws driven into countersunk pilot holes (**see Photo B**).

3 Stand the four legs on the floor and set the bottom panel on top of them. Align the legs with the corners of the bottom panel so their "narrow" strips face the ends of the panel. Drive 2" countersunk screws through the bottom panel and into the ends of the legs to attach them (**see Photo C**).

Photo A: Measure, rip-cut, and crosscut the bottom panel.

Photo B: Use glue and 2" screws to assemble the four legs.

Photo C: Before attaching the legs to the bottom panel, ensure that the legs are all aligned correctly.

Adding the Shelf, Divider, and Top

4 Cut the shelf panel to size and prepare four 13¼"-long spacers about 3" wide from scrap material. Insert a spacer inside each leg and stand it on end on the bottom panel. Clamp the spacers temporarily to the legs so you can rest the shelf panel into place inside the cart assembly. Drive 2" countersunk screws through the faces of the legs into the shelf to secure it, and remove the spacers (**see Photo D**).

5 Cut the divider panel to size, and insert it between the bottom panel and the shelf. Adjust the divider so it's centered on the length of the panels and parallel with their ends. Attach the parts with countersunk 2" screws driven down through the shelf and up through the bottom panel into the divider (**see Photo E**).

6 Cut a panel for the cart's top and set it in position on the cart so its ends and one long edge are flush with the outside faces of the legs. Drill countersunk pilot holes down through the top and into the legs so you can fasten the top to the legs with 2" screws.

7 Invert the cart on the floor, workbench, or a set of sawhorses at a comfortable working height. Rip-cut and crosscut the overhang lamination workpiece to size. Glue and clamp it to the bottom face of the top panel in the overhang area with its edges and ends aligned. Reinforce the glue joint with 1¼" brad nails, if you wish (**see Photo F**).

Photo D: Using clamps and four spacers while securing the shelf to the legs ensures that the shelf is level.

Photo E: Measure and mark a centerline on the bottom and shelf panels to be sure the divider is installed properly.

Photo F: Use glue and clamps to help keep the overhang in place while reinforcing with 1¼" brad nails.

Adding the Side Build-Up Strips

8 While the glue sets, cut a pair of blanks for the side build-up strips. Mark and cut two ¾" x 2½" notches along one long edge of each build-up strip with a jigsaw or band saw so the strips will be able to wrap around the legs when installed (**see Photo G**).

9 Cut the front build-up strip to size. Spread glue over one face of each of the three build-up strips and fit them into place on the top panel inside the cart. Secure the build-up strips with brad nails or 1¼" countersunk wood screws (**see Photo H**).

Adding the Blocking Pieces and Skirts

10 Cut two side blocking pieces and a front and back blocking piece to size. Install them on the bottom face of the bottom panel so the edges of the blocking are flush with the edges of the bottom panel.

11 Position the four casters on the bottom blocking so their mounting plates are flush with the outside corners of the blocking. Attach the casters with ¼" x 1" lag screws and washers driven into the blocking (**see Photo I**).

Photo G: To allow the build-up strips to wrap around the legs, cut two notches along the long edge of each strip.

Photo H: Use glue and brad nails or 1¼" countersunk wood screws to secure the build-up strips to the top panel inside the cart.

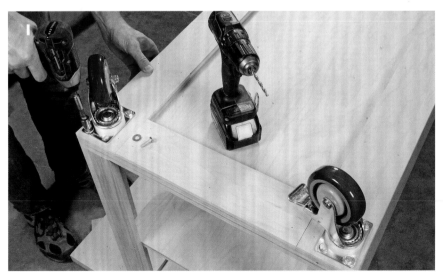

Photo I: Make sure the casters are flush with the outside corners of the blocking before attaching them.

12 Rip-cut and crosscut two side skirt boards and a front and back skirt board from solid lumber. Use a file, sanding block, or a sander to round the sharp corners on the ends of one face of each long skirt board; this will prevent the skirt boards from chipping at the corners during use.

13 Attach the side skirt boards to the narrow ends of the cart's bottom with glue and screws or 2" brad nails. Make sure the ends of the skirt boards are flush with the long edges of the cart bottom. Then install the long skirt boards flush with the faces of the cart bottom and blocking and so the long skirt boards overlap the ends of the short skirt boards. Glue and screw or nail the long skirt boards in place (**see Photo J**).

Photo J: The skirt boards should all be flush with the cart bottom for a proper fit.

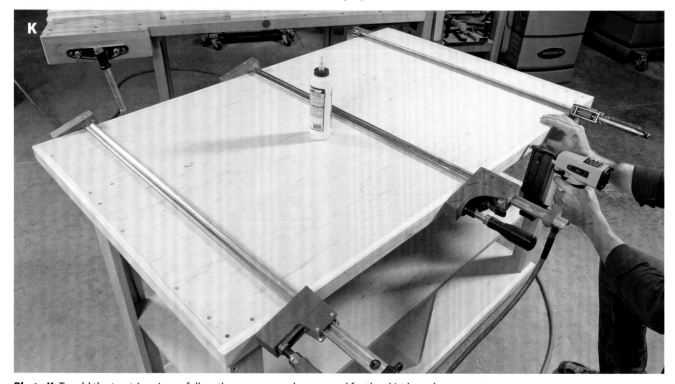

Photo K: To add the top trim pieces, follow the same procedure as used for the skirt boards.

Finishing the Cart

14 Set the cart upright on its casters. Rip-cut and crosscut the side, front, and back top trim pieces from solid lumber. Round the sharp corners on the ends of one face of the front and back trim, just as you did for the front and back skirt boards. Fasten the trim pieces around the edges and ends of the cart top with glue and screws or nails (**see Photo K**).

15 Cover the ends and edges of the shelf with more solid-wood trim cut to size, glued, and screwed or nailed to the shelf panel. In addition to hiding the shelf's edge plywood pieces, the trim pieces also help stiffen the plywood panel. Install the top edges of these trim pieces flush with the top face of the shelf.

16 Cut and install a pair of trim pieces onto the exposed edges of the divider. Center these workpieces on the divider's thickness so they overhang evenly. Attach the trim with glue and screws or nails.

17 Ease the top edges of the cart with a ⅛" or ¼"-radius roundover bit in a handheld router to prevent splinters and to keep workpieces from catching these edges as they exit the saw during cutting (**see Photo L**).

18 Fill the screw and nail head recesses with wood putty or wood plugs if you wish. Applying wood finish to the project is also an option but not necessary on a shop project like this.

Photo L: Easing the top edges with a roundover bit will ensure that your table will be smooth and functional while working.

BASE CABINET

The appeal of custom cabinetry can be yours at a fraction of the price by building this attractive cabinet base. Our simple design features a plastic laminate countertop with solid-oak edging, frame-and-panel doors, a white melamine interior, adjustable shelf, and two drawers that ride on full-extension drawer slides. This base cabinet has finished veneer sides, so it could serve as a stand-alone unit for the kitchen, pantry, laundry room, or even the workshop. You could also modify the design and build as many units as you need to outfit a whole kitchen. If this sounds appealing, see pages 230–239 (and photo, below right) for plans on building a matching oak wall-hung cabinet with glass doors.

See pages 230–239 to build a matching wall-hung cabinet.

Vital Statistics: Base Cabinet

Type: Base cabinet

Overall Size: 25" D x 32" W x 36" H

Material: Melamine particleboard, red oak plywood, solid red oak, red oak veneer, plastic laminate, CDX plywood

Joinery: Miter, biscuit, dado, dowel joints

Construction Details:

- Durable plastic laminate countertop surface
- Frame-and-panel doors
- Cabinet sides laminated with adhesive-backed veneer
- Full-extension metal drawer slides

Finish: Golden oak stain, two coats satin polyurethane varnish

Building Time

Preparing Stock: 3–4 hours

Layout: 2–3 hours

Cutting Parts: 2–3 hours

Assembly: 4–5 hours

Finishing: 1–2 hours

Total: 12–17 hours

Tools

- Table saw
- Power miter saw (optional)
- Drill/driver
- Circular saw
- Biscuit joiner
- Iron
- Veneer edge trimmer
- J-roller
- Router table
- Router with flush-trimming bit, ¼" straight bit, chamfer bit
- Clamps
- Doweling jig

Shopping List

- ⭘ ¼" x 48" x 48" melamine
- ⭘ ½" x 24" x 48" melamine
- ⭘ ¾" x 4' x 8' melamine
- ⭘ ¾" x 6" x 3' red oak
- ⭘ ¾" x 2' x 4' CDX plywood
- ⭘ ¼" x 24" x 24" oak plywood
- ⭘ 1⁄32" x 2' x 3' laminate
- ⭘ 1⁄32" x 2' x 6' adhesive-backed red oak veneer
- ⭘ Red oak edge banding (15')
- ⭘ PVC white edge banding (15')
- ⭘ (4) 2" x 1½" ball-tip hinges
- ⭘ (4) ¼"-dia. brass shelf pins
- ⭘ (4) 1³⁄16"-dia. brass knobs
- ⭘ (4) 20" drawer slides
- ⭘ Magnetic door latch
- ⭘ Biscuits, ⅜"-dia. dowel
- ⭘ Contact cement

Exploded View

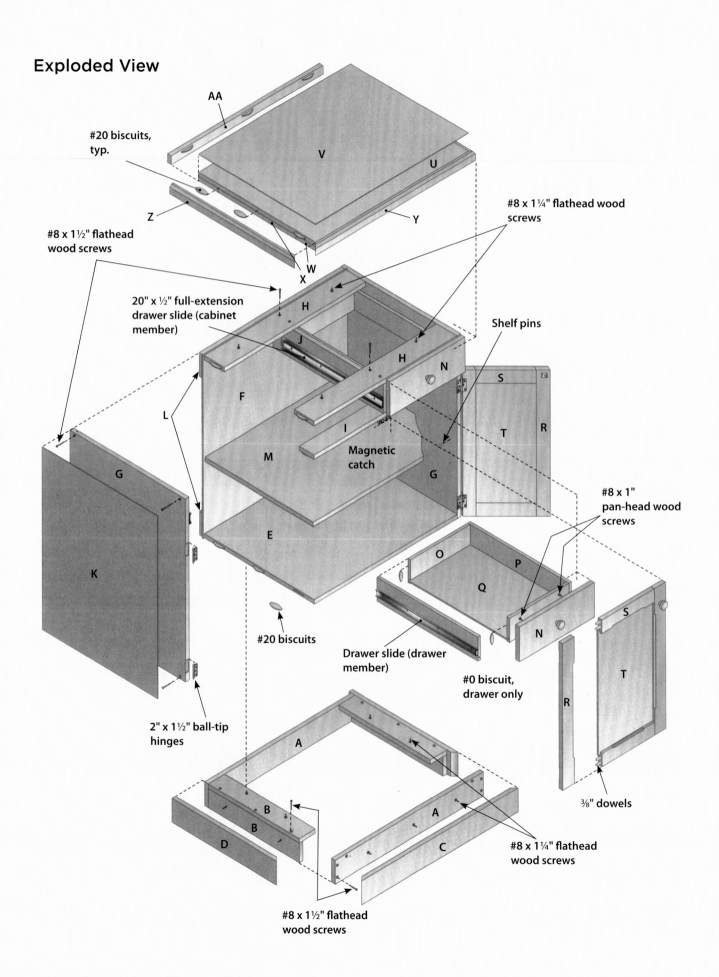

AA

#20 biscuits, typ.

#8 x 1½" flathead wood screws

Z

Y

V

U

W

X

#8 x 1¼" flathead wood screws

20" x ½" full-extension drawer slide (cabinet member)

H

J

H

Shelf pins

N

S

F

T

R

I

Magnetic catch

M

G

#8 x 1"
pan-head wood screws

G

L

E

O

P

K

Q

N

#20 biscuits

Drawer slide (drawer member)

S

#0 biscuit, drawer only

R

T

2" x 1½" ball-tip hinges

⅜" dowels

A

B

B

A

B

D

C

#8 x 1¼" flathead wood screws

#8 x 1½" flathead wood screws

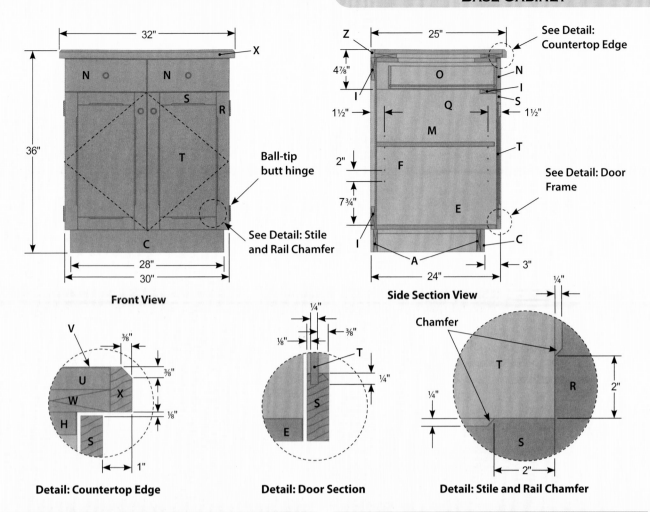

Front View

Side Section View

Detail: Countertop Edge

Detail: Door Section

Detail: Stile and Rail Chamfer

Cabinet Cutting List

	Part	No.	Size	Material
A.	Base front/back	2	¾" x 4" x 26½"	CDX plywood
B.	Base stretcher	4	¾" x 3¼" x 18¼"	"
C.	Front toekick	1	¾" x 4" x 28"	Red oak
D.	Side toekick	2	¾" x 4" x 21"	"
E.	Bottom	1	¾" x 22¼" x 28½"	Melamine
F.	Back	1	¼" x 29" x 30½"	"
G.	Side	2	¾" x 23½" x 30½"	"
H.	Top stretcher	2	¾" x 4" x 28½"	"
I.	Drawer rail	1	¾" x 3" x 28½"	"
J.	Drawer divider	1	¾" x 4⅞" x 22¼"	"
K.	Veneer face	2	1/32" x 24" x 32"	Oak veneer
L.	Back blocking	2	¾" x 4" x 28½"	CDX plywood
M.	Shelf	1	¾" x 21¾" x 28¼"	Melamine
N.	Face	2	¾" x 5³⁄₁₆" x 14⁵⁄₁₆"	Red oak
O.	Front/back	4	½" x 4" x 11¹⁄₁₆"	Melamine

Drawers Cutting List

	Part	No.	Size	Material
P.	Side	4	½" x 4" x 20"	Melamine
Q.	Bottom	2	¼" x 12¹⁄₁₆" x 19½"	"
R.	Stile	4	¾" x 2¼" x 24⁷⁄₁₆"	Red oak
S.	Rail	4	¾" x 2¼" x 10⁷⁄₁₆"	"

Door Cutting List

	Part	No.	Size	Material
T.	Panel	2	¼" x 10¹⁵⁄₁₆" x 20¹¹⁄₁₆"	Oak plywood
U.	Core	1	¾" x 23½" x 30½"	Melamine
V.	Top	1	1/32" x 24" x 32"	Laminate
W.	Build-up	2	¾" x 4" x 15½"	Melamine

Counter Cutting List

	Part	No.	Size	Material
X.	Build-up	2	¾" x 4" x 30½"	Melamine
Y.	Front edge	1	¾" x 1½" x 32"	Red oak
Z.	Side edge	2	¾" x 1½" x 25"	"
AA.	Back edge	1	¾" x 1½" x 30½"	"

Base Cabinet: Step-by-Step

Construct the Base

1 Rip-cut and crosscut the base front, back, and four stretchers to size from ¾" CDX plywood. Attach the face of one stretcher to the long edge of another stretcher with glue and 1½" flathead wood screws, forming an L-shaped assembly. Build another L-shaped assembly with the other two stretchers. Place the assembled stretchers between the base front and back pieces, keeping the outside surfaces of the stretchers flush with the ends of the front and back. Drill countersunk pilot holes and drive 1½" flathead wood screws through the faces of the base front and back and into the ends of the stretchers.

2 Cut the front and side toe kicks to size from solid red oak. Bevel one end of each side toe kick and both ends of the front toe kick so the pieces will fit around the assembled base with no visible end grain. Make these miter cuts on a power miter saw or on the table saw with the blade tilted to 45°. Set the toe kick parts in position and mark the back ends of the side toe kicks so they'll extend ½" beyond the back of the base. Crosscut the side toe kicks to length. Turn the entire base unit upside down and attach the toe kick parts to the base with glue and countersunk 1¼" screws, driven from inside the base (**see Photo A**). Be sure the miters meet snugly at the corners and the top and bottom edges are flush all around.

Build the Cabinet Carcase

3 Cut the cabinet sides, top stretchers, bottom, drawer rail, drawer divider, and adjustable shelf to size from ¾" white melamine-covered particleboard.

4 Lay out and drill ¼"-diameter shelf pin holes in the sides, spaced vertically, 2" on-center and 1½" in from each edge. We started the holes 8½" up from the bottom ends of the cabinet sides. The holes must align on both cabinet sides, or the shelves won't be level when the cabinet is assembled. To help align the holes, make a drilling template using pegboard, and clamp it in the same position when you lay out the holes on each cabinet side. Drill the shelf pin holes ⅜" deep (a depth stop installed on your drill bit will keep you from accidentally drilling through the cabinet sides).

5 Cut a ¼"-wide, ⁵⁄₁₆"-deep dado, ¹³⁄₁₆" from the back inside edge of each cabinet side. The cabinet back will slide into these dadoes later. Since sheet material often does not match the nominal thickness precisely, measure the thickness of the panel you'll use for the cabinet back and adjust the width of the dadoes accordingly so your cabinet back will be sure to fit.

Photo A: Glue and screw the toekick parts to the base assembly. Hold the miters together tightly as you drive screws from inside the base.

Note

Melamine is particularly prone to tearout or chipping. If you cut these parts on a table saw or with a circular saw, install a plywood blade, triple-chip blade, or combination blade set low to minimize chipping the melamine. For added insurance, run a strip of masking tape along your cutting lines before you cut the parts. Masking tape will help keep the melamine edges intact while you saw. Remove the tape once the parts are cut.

6 Lay out and cut slots for #20 biscuits in the cabinet sides, top stretchers, drawer rail, and bottom to help line up the parts during assembly. The cabinet sides overlap the ends of the bottom and stretchers. Lay out the cabinet parts so the back edges of the bottom and rear top stretcher are flush with the front edge of the cabinet back dadoes. The front edges of the front top stretcher and the cabinet bottom should be flush with the front edges of the cabinet sides. Glue biscuits into the ends of the top stretchers, drawer rail, and bottom. Wipe off excess glue with a damp cloth. Dry-fit the assembly and hold it together with clamps. Drill countersunk pilot holes for 1½" flathead wood screws through the cabinet sides and into the horizontal cabinet parts. Disassemble the cabinet and spread glue into the remaining biscuit slots and mating joint surfaces. Reassemble and clamp up the cabinet, checking to make sure it is square. Reinforce the joints with screws (**see Photo B**). Measure and mark the drawer divider location, centering it on the cabinet width. Attach the divider with 1½" countersunk screws driven down through the top stretchers and up through the drawer rail.

7 Apply red oak iron-on veneer tape to the front edges of the cabinet sides, bottom, top front stretcher, drawer rail, and the front end of the drawer divider. Trim off overhanging tape with a hand edging trimmer, or hold a sharp chisel flat against the panel and push it slowly along to slice away the excess.

8 Cut the cabinet back to size from ¼"-thick white melamine board. Slide the cabinet back into the dadoes on the cabinet sides so the melamine side faces into the cabinet. Check the cabinet for square by measuring diagonally from the cabinet bottom to the top in both directions. When the measurements match, the cabinet is square. Attach the back with countersunk 1¼" screws driven through the back and into the cabinet bottom and top rear stretcher.

9 Cut the two back blocking pieces to size. Glue and screw the blocking flush against the back edges of the cabinet sides, at the top and bottom of the carcase, using countersunk 1½" screws. Since the cabinet sides will be covered with veneer, attach the blocking by driving the screws through the cabinet sides behind the back panel dadoes and into the ends of the back blocking. The back blocking pieces serve as fastening cleats when you screw the cabinet to a wall. Fill the screw holes in the cabinet sides with wood putty and sand smooth.

Attach the Sheet Veneer

10 Cover the cabinet sides with sheet veneer (we used adhesive-backed cabinet veneer). Cut the veneer to size according to the dimensions given in the cutting list on page 223, using scissors or a utility knife against a steel straightedge. The veneer sheets are oversized at this stage to allow some leeway in positioning them, since the adhesive grabs on contact.

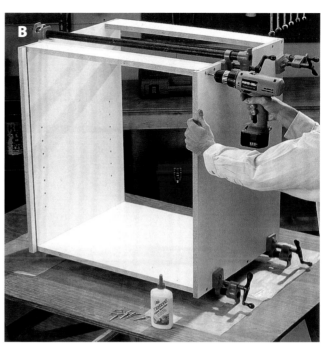

Photo B: Assemble the cabinet carcase. Cut biscuit slots to align the parts, then glue and clamp the carcase and fasten the joints with screws. Drill countersunk pilot holes before driving the screws.

Tip

If you do get an air bubble trapped beneath the veneer, push a stick pin through the veneer or make a tiny slit with a thin craft knife. This perforation should allow the air to escape so you can roll or press the veneer flat.

11 Scuff the outer melamine faces of the cabinet sides with 80-grit sandpaper to de-gloss the melamine, which improves the bonding surface for the veneer adhesive. Lay the cabinet on its side on strips of scrap plywood. Place one of the veneer sheets on the cabinet side, adhesive side down, and center the veneer so there is an even overhang all around. Hold it in place by clamping a couple spring clamps along one edge. Peel back the backing paper a few inches from the opposite edge. Press this edge down firmly to bond the adhesive, then remove the spring clamps and begin to slowly pull the paper off while you press the veneer down flat with a J-roller or scrap of 2 x 4 wrapped in a towel. Smooth the veneer using overlapping strokes all the way across the sheet as you remove the paper. Try not to trap air bubbles. Proceed with this peeling and smoothing process until all the paper is removed and you have completely flattened the veneer.

12 Trim away the excess veneer. Use a router with a piloted flush-trimming bit to trim the overhanging veneer all around (**see Photo C**). Turn the cabinet over and veneer the other cabinet side.

Make the Drawers

13 Prepare the drawer box parts. Cut the drawer fronts, backs, and sides to size from ½" melamine particleboard. Cut the drawer bottoms from ¼" melamine. Use a table saw and dado-blade set to machine a ¼"-wide, ⁵⁄₁₆"-deep dado ¼" from one long edge into the drawer fronts, backs, and sides (**see Photo D**). The dadoes will house the drawer bottoms. Conceal the top edges of the drawer fronts, backs, and sides with white iron-on edge tape, and trim the tape flush.

14 Build the drawer boxes. Lay out and cut centered #0 biscuit joints for assembling the drawers. Lay out the biscuit joints so the drawer front and back panels fit between the sides. Glue and clamp up the drawers with the drawer bottoms in place (**see Photo E**).

15 Install the drawer faces. Cut the drawer faces to size from ¾" solid red oak. Attach them to the fronts of the drawers so that the faces overhang the bottom edges of the drawer box by ½". The inside end of each drawer front should overhang the drawer side by ½", and the outside end should be flush with the outside face of the cabinet side. Drill pilot holes through the insides of the drawers and into the backs of the drawer faces. It can be helpful to enlarge the holes from side to side in the drawer fronts so that you can align the faces after the drawers are hung. Screw the drawer faces to the drawer fronts, using two 1" pan-head screws per drawer (**see Photo F**).

Photo C: After you apply adhesive-backed veneer to one cabinet side, trim off overhanging veneer with a router, and piloted flush-trimming bit. Apply veneer to the other cabinet side using the same method.

Photo D: Cut dadoes for the drawer bottoms in the drawer fronts, backs, and sides. Adjust the dado width to match the thickness of the drawer bottom panels you buy—panel thickness may vary.

Photo E: Dry-fit the drawer box parts, then glue and clamp up each drawer, with the drawer bottom inserted in its dado. Use #0 biscuits to align the joints. Wax paper protects your worksurface from excess glue. Check the drawers for square and adjust the clamps as needed.

Photo F: Use spring clamps to hold the drawer face in position on the drawer while you drill pilot holes and insert screws. Because the drawer front overhangs the bottom of the drawer by ½", shim the back of the drawer above the worksurface with ½" scrap plywood.

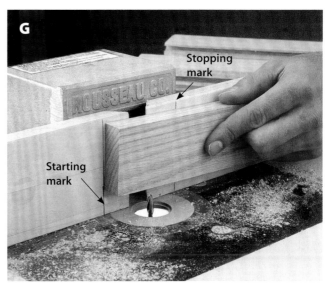

Photo G: Draw reference marks on the router table fence that index the starting and stopping points for cutting the stopped dadoes in the door stiles. Start each stopped dado by aligning one end of a stile with the starting mark on the fence and lowering the workpiece onto the bit.

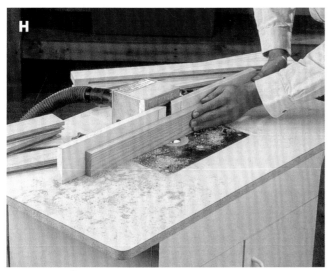

Photo H: Once the bit is buried in the workpiece at the starting mark, slide each door stile along the fence until the other end of the workpiece reaches the stop mark. Turn off the router and lift up the board to finish the cut.

Build and Hang the Doors

16 Cut the door stiles and rails to size from solid red oak stock and flatten the edges on a jointer or with a hand plane. The edges must be flat so the rails and stiles will fit together tightly where the parts meet in the corners. Cut the door panels from ¼" red oak plywood.

17 The door panels "float" in dadoes cut on the inside edges of the door stiles and rails. The dadoes in the stiles must stop 2" from the ends, so cut the dadoes on a router table with starting and stopping marks laid out on the fence to guide the cuts (**see Photo G**). Install a ¼"-wide straight bit and raise the bit to ⁷⁄₁₆" above the table. Set the router fence ⅛" away from the bit. This way, the dadoes will be offset across the width of the doorframe parts to allow for a ⅜" reveal between the door panel and the face of the doorframe.

18 Cut the door stile dadoes. To make the stopped dadoes, start the router, align the stile with the starting mark on the router fence, and lower it slowly down onto the bit (**see Photo H**). Slide the stile along the fence until the other end of the stile reaches the stop mark on the fence. Shut off the router and remove the workpiece. Since the dadoes in the rails run the full length of the inside edge, rout all the way along one edge and through both ends of each rail.

19 Rout a ¼" x ¼" chamfer along the inside front edge of each door stile and rail. Stop the chamfers 2" from the ends of the rails and 4¼" from the ends of the stiles. Use a chamfering bit in the router table, and mark starting and stopping points on the router fence to make these stopped cuts.

Photo I: Stain the door panels, then glue and clamp up the doors using ⅜"-diameter dowels to reinforce the rail and stile corner joints.

Photo J: Attach solid-oak edging pieces around the laminate countertop, using biscuits and glue. The edging fits flush with the countertop.

20 Drill ⅜"-diameter dowel joints with a doweling jig to join the rails to the stiles. Position the dowel holes on the mating surfaces of the rails and stiles, ½" and 1½" from the top and bottom edges of the door. Drill the holes ⅞" deep in all the parts.

21 If you plan to stain this project, stain the door panels before assembling the doors, since you'll still have full access to the door panels. We used golden oak stain.

22 Assemble the doors. Dry-assemble the rails, stiles, and panels first to check the fit of the parts. Disassemble the doors and spread glue into the dowel holes and mating surfaces of the joints. Insert 1½"-long dowels and clamp up both doors (**see Photo I**). Check for square and clean up glue squeeze-out before the glue dries.

23 Lay out and cut mortises on the front edge of the cabinet sides and the back face of the stiles for ball-tip hinges.

Make the Countertop

24 Cut the countertop core and four build-up pieces to size from ¾" melamine particleboard. Glue and screw the build-up to the underside of the core with 1¼" screws.

25 Cut the laminate to the oversized dimensions given in the cutting list. Use a laminate scoring knife (score the laminate several times against a steel straightedge, then break it along the scored cut) or cut the laminate on the table saw using a laminate-cutting blade. Scuff the glossy melamine surface of the core with 80-grit sandpaper. Apply contact cement to both the laminate and the top core, using a paint roller with a foam sleeve. After the contact cement becomes tacky, set the laminate in place on the core. Be careful here. The cement bonds instantly on contact, making it extremely difficult to remove or reposition the laminate. To make this process more foolproof, lay several thin plywood strips across the width of the core, say every 6" or so. Lay the laminate on top of the strips and align the laminate over the core. Starting at one end, remove one stick at a time and press down the laminate with a J-roller. Once you've removed all the sticks, roll the laminate down firmly all over the surface with the J-roller to ensure a good bond. Trim the laminate to final size with a router and flush-trimming bit. (See page 189 for more information on applying laminate.)

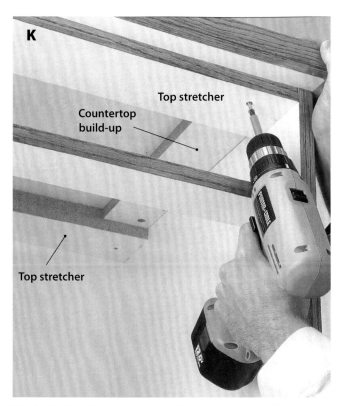

Photo K: Attach the countertop to the top stretchers with countersunk 1½" screws, driven from inside the cabinet, up through the stretchers, and into the countertop build-up.

Photo L: Hang the doors and install door and drawer knobs. Loosen the drawer face screws a bit and adjust the drawer faces so they fit evenly above the doors and are flush with the cabinet sides. Tighten the screws.

26 Rip-cut and crosscut the front, side, and back countertop edging from solid red oak. Fit these edges around the core, mitering the front corners and butting the rear corners. Cut slots for #20 biscuits in the edging and core to help align the parts. Glue and clamp the edging to the core (**see Photo J**). When the glue dries, remove the clamps and ease the edges and corners of the side and front edging by routing a ⅜" chamfer. Rout the chamfer on the front edging first, then rout the side edging (doing this will keep the bit from tearing out the corners).

Finishing Touches

27 Sand all the red oak surfaces and ease sharp edges with 150-grit sandpaper. Apply the finish of your choice. We finished our cabinet with golden oak stain to match the door panels and applied two coats of satin polyurethane varnish.

28 Attach the base to the cabinet with countersunk 1¼" screws driven up from under the base. Assemble the parts so the back of the base is flush with the back of the cabinet and the cabinet overhangs the base evenly on both sides.

29 Attach the countertop to the cabinet with countersunk 1½" screws driven through the top stretchers into the countertop build-up (**see Photo K**). The countertop should be flush with the back of the cabinet and overhang the cabinet sides equally.

30 Attach the metal drawer slide hardware to the drawer boxes, drawer divider, and inside cabinet faces according to the manufacturer's instructions. Hang the drawers. Insert the shelf pins into the holes and set the shelf in place. Attach the hinges to the doors and cabinet carcase, and hang the doors. Loosen the drawer face screws just enough to adjust the alignment of the drawer faces so they're centered over the doors, then tighten the screws. Install the knobs on the doors and drawers (**see Photo L**).

WALL-HUNG CABINET

Create attractive storage space for displaying glassware, china, or collectibles with this oak wall-hung cabinet. The shelves are adjustable to accommodate items of varying sizes, and the cabinet mounts to a wall with hidden, interlocking French cleats. We've styled this cabinet with glass doors to function as a companion piece for the base cabinet project found on pages 220–229.

Vital Statistics: Wall-Hung Cabinet

Type: Wall-hung cabinet

Overall Size: 15" D x 32" W x 31½" H

Material: Red oak plywood, solid red oak, glass

Joinery: Dowel, miter, biscuit, dado joints

Construction Details:

- Solid-oak doorframes
- Glass mounted in door rabbets and held in place with wooden glass stops
- Iron-on oak veneer edge tape
- Cabinet mounted with concealed French cleats
- Two adjustable shelves

Finish: Golden oak stain; two coats satin polyurethane varnish

Building Time

 Preparing Stock: 4–5 hours

 Layout: 3–4 hours

 Cutting Parts: 2–3 hours

 Assembly: 6–8 hours

 Finishing: 1–2 hours

Total: 16–22 hours

Tools You'll Use

- Table saw
- Power miter saw (optional)
- Drill/driver
- Doweling jig
- Biscuit joiner
- Clamps
- Router with ⅜" rabbeting bit, piloted chamfer bit, ¼" roundover bit
- Chisel
- Marking knife
- Brad pusher
- Nailset

Shopping list

- ○ ¾" x 4' x 8' red oak plywood
- ○ ¼" x 4' x 8' red oak plywood
- ○ ¾" x 4" solid red oak (16 lineal')
- ○ (2) ⅛" x 10 $^{15}⁄_{16}$" x 26⅛" tempered glass
- ○ Red oak edge banding (20')
- ○ (4) 2" x 1½" ball-tip hinges
- ○ (8) ¼"-dia. shelf pins
- ○ (2) Brass doorknobs
- ○ (2) Magnetic door catches
- ○ #8 x 2" flathead wood screws. ¾" wire brads
- ○ #20 biscuits, ⅜"-dia. dowel
- ○ Wood glue
- ○ Finishing materials

Exploded View

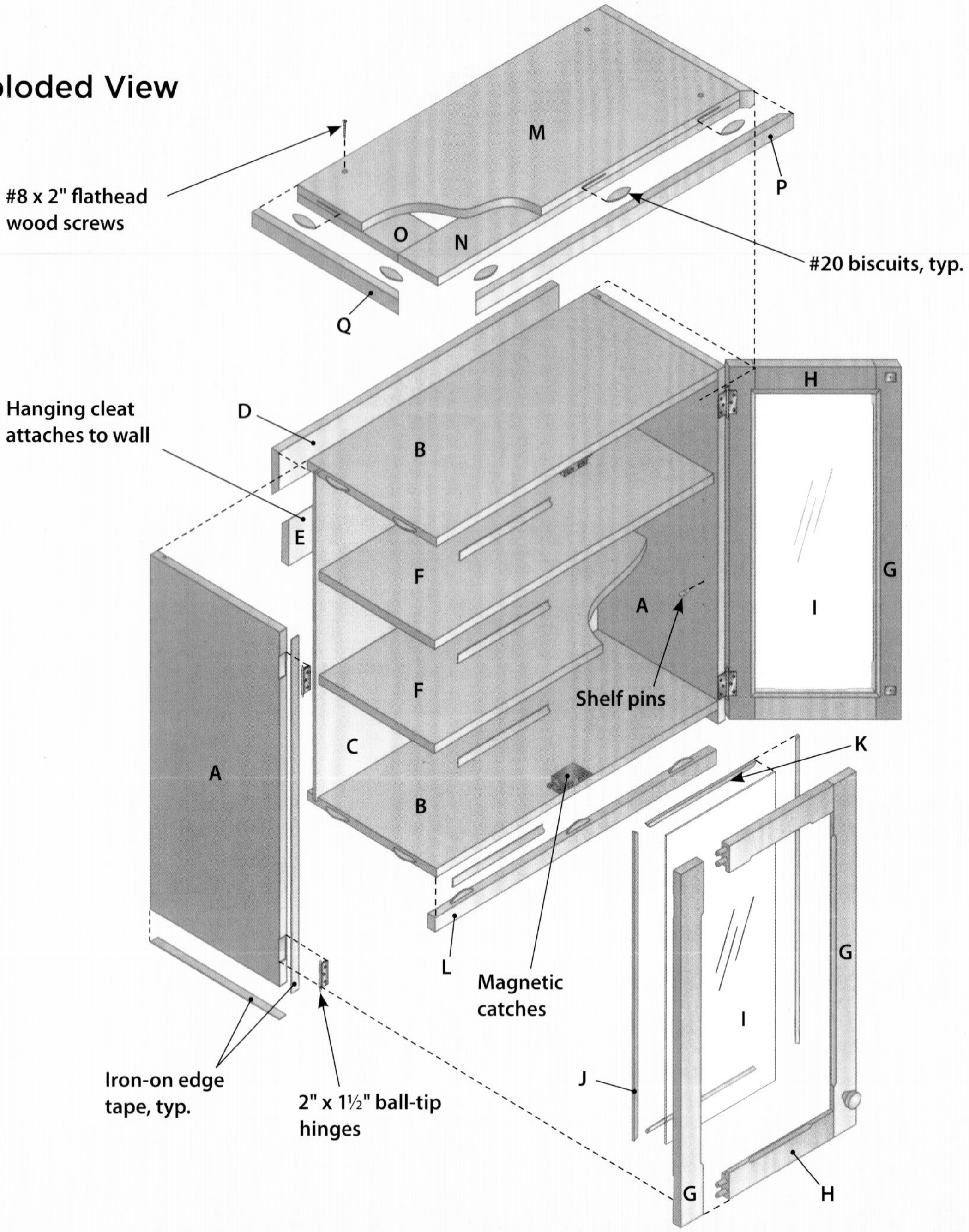

#8 x 2" flathead wood screws
M
P
O
N
#20 biscuits, typ.
Q
Hanging cleat attaches to wall
D
B
H
E
F
A
I
G
F
Shelf pins
A
C
K
B
L
Magnetic catches
G
I
Iron-on edge tape, typ.
2" x 1½" ball-tip hinges
J
G
H

Wall-hung Cutting List							
Part	**No.**	**Size**	**Material**	**Part**	**No.**	**Size**	**Material**
A. Side	2	¾" x 13⅛" x 30"	Red oak plywood	**J.** Glass stop	4	¼" x ⅜" x 26⅞"	Red oak
B. Top/bottom	2	¾" x 13⅛" x 28½"	"	**K.** Glass stop	4	¼" x ⅜" x 10⁷⁄₁₆"	"
C. Back	1	¼" x 29" x 28"	"	**L.** Light valance	1	¾" x 1¼" x 28½"	"
D. Cabinet cleat	1	¾" x 4" x 28½"	"	**M.** Crown	1	¾" x 14¼" x 30½"	Red oak plywood
E. Wall cleat	1	¾" x 4" x 26½"	"	**N.** Build-up	1	¾" x 4" x 30½"	"
F. Shelf	2	¾" x 11¾" x 28½"	"	**O.** Build-up	2	¾" x 4" x 10¼"	"
G. Door stile	4	¾" x 2¼" x 29⅞"	Red oak	**P.** Front edge	1	¾" x 1½" x 32"	Red oak
H. Door rail	4	¾" x 2¼" x 10⁷⁄₁₆"	"	**Q.** Side edge	2	¾" x 1½" x 15"	"
I. Glass panel	2	⅛" x 10 ¹⁵⁄₁₆" x 26⅛"	Tempered glass				

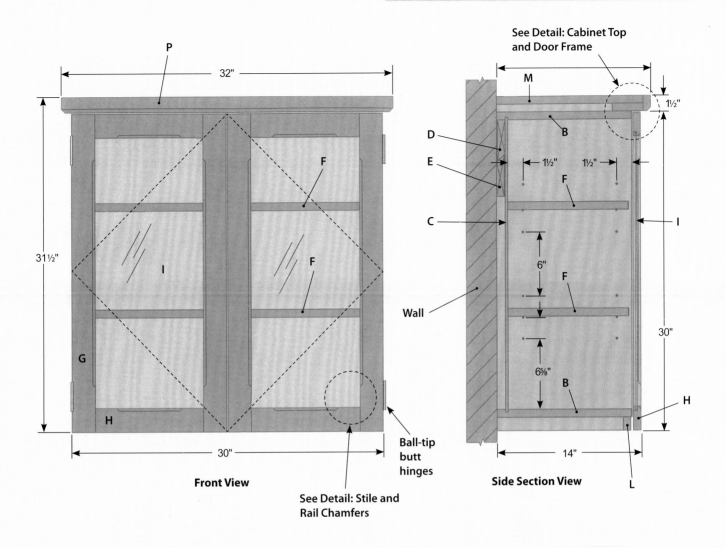

See Detail: Cabinet Top and Door Frame

P

32"

31½"

F

I

F

G

H

30"

Front View

Ball-tip butt hinges

See Detail: Stile and Rail Chamfers

M

B

D

E

C

Wall

1½" 1½"

F

6"

F

6⅝"

B

14"

30"

1½"

I

H

L

Side Section View

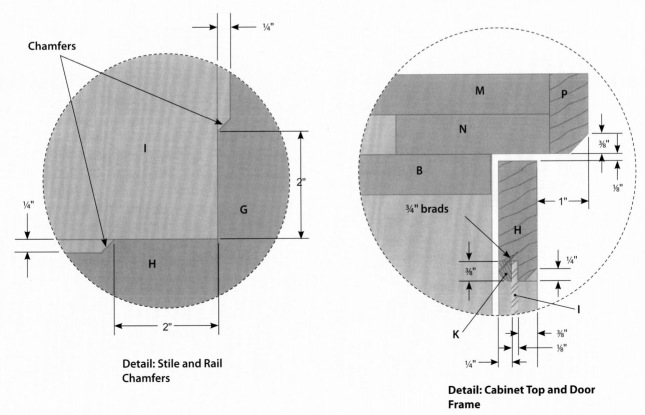

Chamfers

¼"

I

2"

G

¼"

H

2"

Detail: Stile and Rail Chamfers

M

P

N

⅜"

B

⅛"

1"

¾" brads

H

¼"

⅜"

K

I

¼" ⅜" ⅛"

Detail: Cabinet Top and Door Frame

Wall-Hung Cabinet: Step-by-Step

Build the Carcase

1 Cut the cabinet sides, top, bottom, and shelves to size from ¾" red oak plywood, according to the dimensions given in the cutting list, page 232. Use a sharp plywood or combination table saw blade to minimize chipping out the veneer on the underside of the panels, especially during crosscutting.

2 Lay out and drill ¼"-diameter holes for the adjustable shelf pins, spaced at 2" intervals on the inside faces of the cabinet sides. Position the rows of holes 1½" in from the front and back edges of the side panels. Use a piece of pegboard as a template for drilling the holes. Be sure to set up the template exactly the same way on each cabinet side so the shelf pin holes will line up once the carcase is assembled.

3 Cut a ¼"-wide, ⁵⁄₁₆"-deep dado along the inside faces of the cabinet sides, top, and bottom, ¹³⁄₁₆" from the back edge of each part (**see Photo A**). Cut the dadoes on a table saw with a dado-blade set. The dadoes will house the cabinet back. Since ¼" plywood is almost never exactly ¼" thick, measure the thickness of the panel stock you'll use for the back before setting up the dado blade. This way you can be sure the back will fit the dadoes.

4 Rip-cut and crosscut the cabinet back to size from ¼" red oak plywood.

5 Fasten iron-on red oak veneer edge tape to the front edges of the cabinet sides, top, bottom, and adjustable shelves (see page 186). Trim off overhanging tape with a hand edge-trimming tool or use a sharp chisel, holding it flat against the panel as you run it carefully along the edges. Then apply veneer tape to the bottom edges of the cabinet sides.

6 Lay out and cut slots for #20 biscuits in the ends of the cabinet top and bottom, and in the inside faces of the sides to help align these parts during assembly. Note that the cabinet bottom is located 1¼" from the bottom edges of the sides. Glue biscuits into the ends of the top and bottom, and wipe away excess glue with a rag.

Photo A: Cut a ¼"-wide, ⅝"-deep dado, ¹³⁄₁₆" in from the back edge of each of the cabinet sides, top, and bottom. Adjust the exact dado width to match the actual thickness of the cabinet back panel.

7 Dry-fit the carcase parts (sides, top, bottom, and back) to test the fit, then disassemble the cabinet. Spread glue on the mating parts and into the biscuit slots on the cabinet sides, and reassemble the carcase. Clamp up the cabinet, making sure it is square and adjusting the clamps as necessary before the glue dries (**see Photo B**). Clean up glue squeeze-out.

8 Cut the light valance strip to size from solid red oak. Cut slots for #20 biscuits in the top edge of the valance and in the underside of the cabinet bottom. Glue and clamp the valance in place.

9 Cut the cabinet cleat and the wall cleat to size, bevel-ripping one edge of each cleat at 45°. Glue and clamp the cabinet cleat in place, butting it up against the bottom edge of the cabinet top and flat against the back panel (see Side Section View, page 233). The beveled edge should face down and toward the cabinet back panel so it will interlock with the wall cleat when you hang the cabinet.

Build the Doors

10 Cut the door stiles and rails to size from solid red oak stock. It is important that the stile and rail stock be straight and flat, especially in cases when the door won't be stiffened with a wood center panel. Stock that isn't flat and straight will produce warped or twisted doors that won't sit flat against the cabinet and will be uneven. If your red oak stock is less than ideal, a better alternative might be to mill your door stock from thicker oak lumber (see sidebar, below).

Photo B: Glue up the cabinet carcase with the back in place. Measure across the diagonals to make sure the assembly is square, and adjust the clamps as needed. Use scrap wood blocks to pad the clamps and protect the cabinet sides.

Milling Flat Stock from Thicker Lumber

Finding flat stock can be difficult when you are buying ¾"-thick boards planed smooth on two faces. This is because wood typically is not jointed flat before it is planed to final thickness at the mill, and any internal stresses in the wood cause it to warp again. If you can't find suitable surfaced lumber, you may need to joint and plane your ¾" stock from thicker 4/4 boards (about 1" nominal thickness). First joint the faces of 4/4 stock to flatten them. If the lumber was warped or twisted when you bought it, let it sit for a few days, then joint the board faces flat again. Once you've corrected for warp or twist by repeated passes over a jointer, plane the flattened stock down to ¾".

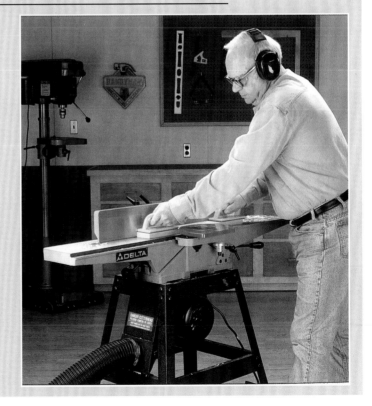

11 Use a doweling jig to drill ⅜"-diameter, ⅞"-deep dowel holes in the ends of the rails and the edges of the stiles for assembling the doorframes. Space the dowel holes ½" and 1¼" from both ends of each stile, then drill corresponding holes in the rails. Spread glue over the mating parts of the joints, insert 1½"-long, ⅜"-diameter dowels, and clamp up the doorframes (**see Photo C**). Make sure the frames are square.

12 Mill a ⅜" x ⅜" rabbet inside the doorframes to house the glass panels. Install a ⅜" piloted rabbeting bit in your router. To rout the rabbets, clamp the doorframes to your worksurface with the inside face up. Set the bit depth to about ¼". Rout a rabbet all around the inside edge of the frame. Then set the bit to the full ⅜" depth and make a second pass to complete each rabbet (**see Photo D**). Since the router will leave rounded corners in the doorframes, square corners with a sharp chisel.

13 Turn the doorframes over with their front faces up. Install a piloted chamfer bit in the router. Rout a ¼" x ¼" chamfer along the inner edges of the rails and stiles. Start and stop the chamfer on each rail and stile 2" from the inside frame corners, as shown in the Detail: Stile and Rail Chamfers drawing, page 233. Clamp temporary stopblocks to the doorframes to index your router base when routing these chamfers.

14 Lay out and mark hinge mortises on the inside faces of the doors. Locate the hinges 2" in from the top and bottom door ends. For greatest accuracy, use a marking knife or sharp utility knife against a straightedge to lay out the mortises, instead of a pencil. Using one of these two marking tools scores the wood grain, making it easier and neater to pare right up to your layout lines with a chisel.

Photo C: Glue and clamp up the door frames with ⅜"-dia. dowel joints. Line your work area with wax paper to protect the benchtop from glue squeeze-out.

Photo D: Use a piloted rabbeting bit to rout a ⅜" x ⅜" rabbet around the inside back edge of each doorframe for the glass panel. Rout the rabbets in two passes.

Photo E: Lay out and cut mortises in the doors for the butt hinges (two per door). Work carefully, especially when chiseling next to your mortise layout lines.

15 Remove the waste in the hinge mortise areas with a chisel (**see Photo E**). The metal hinge leaves should sit flush with each door face once the mortises are cut. To keep the chisel from diving too deeply into the wood as you pare the mortises, hold it with the blade bevel facing down and shear off thin shavings up to your layout lines. You could also cut these hinge mortises with a straight bit in the router set to a depth that matches the hinge leaf thickness. When routing hinge mortises, stop short of your layout lines and clean out the rest of the waste in the mortise area with a chisel.

16 Drill pilot holes for the screws and mount the hinges to the doors. Lay a door into position on each side of the cabinet, and support the doors from beneath with scrap wood blocks. Align the ends of the doors with the top and bottom of the cabinet, and trace around the hinge leaves to mark mortise locations on the front edges of the cabinet (**see Photo F**).

17 Cut hinge mortises in the cabinet with a chisel as before, then drill pilot holes for the hinge screws and install the doors. Test the action of the doors and their relationship to one another when closed. You can make minor door adjustments by loosening the hinge screws and moving the hinges out or in slightly, or by shimming beneath the hinge leaves with strips of paper or masking tape. Remove the doors and hinges.

Photo F: Support the doors next to the cabinet with scrap, and trace around the hinges to mark mortises on the cabinet sides. Cut these mortises with a chisel.

Mill the Glass Stops

18 The glass panels are secured in the door rabbets with wooden stops and ¾" brads. To make the stops, start with a length of ¾"-thick red oak with one edge jointed flat and square. Rout a ¼" roundover along the top and bottom of this edge. Set the table saw rip fence ⅜" from the blade. Lay the board facedown on the saw table with the roundover edge against the rip fence, and rip a ⅜" strip off the board, using a push stick to guide the narrow strip past the blade. Now set the saw fence ¼" from the blade. Lay the roundover strip flat on the saw table and rip a ¼"-wide piece. This ⅜" x ¼" stick, with one rounded edge, forms a length of glass stop. Flip the initial remaining strip around so the other rounded edge is against the saw fence and rip a second stop. Mill all eight stops similarly, then crosscut four glass stops per door and miter the ends so the stops fit tightly in the rabbets.

Build the Cabinet Crown

19 Cut the cabinet crown to size and lay it upside down on the workbench. Cut the side and front buildups to size and fasten them to the crown with 1¼" countersunk wood screws. Center the screws across the width of the build-up pieces (**see Photo G**).

20 Rip the front and side edges to width from ¾" solid red oak, leaving the pieces longer than necessary. Miter the front corners of the front and side edging to fit around the crown using a power miter saw or the table saw. Trim the back ends of the side edging square and flush with the back of the core.

21 Cut slots for #20 biscuits in the front and side edges and the crown to help align the parts. Glue and clamp the edging to the crown (**see Photo H**).

Photo G: Screw the front and side build-ups to the underside of the crown, with the edges flush all around. Center the screws across the width of the build-ups.

Photo H: Glue the solid oak front and side edging to the crown. Miter the front corners and install #20 biscuits to keep the edging aligned. Clamp up the assembly to hold the mitered corners tight while the glue dries.

Photo I: Secure the glass in the door recesses with wooden glass stops. Drive ¾" wire brads at a 45° angle through the stops and into the doorframes. Use a brad pusher (foreground) to avoid hammering close to the glass. Then tap the brad heads flush with a nail set.

22 Use a router with a piloted chamfer bit to rout a ¼" chamfer around the bottom edge of the crown. Rout the chamfer along the front edging piece first, then cut the chamfers along the sides. Cutting the chamfers in this order minimizes tearout on the ends of the front edging piece.

Finishing Touches

23 Sand all inside and outside cabinet surfaces with 120-grit sandpaper, then finish-sand with 150-grit.

24 Apply the finish of your choice. We used golden oak stain and top-coated with two coats of satin polyurethane varnish to match the base cabinet project, pages 220–229. Leave the top of the cabinet and the bottom of the crown free of wood finish.

25 Place the crown on the cabinet top and adjust it so the back of the crown is flush with the back edge of the top and the sides overhang evenly. Drill countersunk screw holes, counterbored for wood plugs, down through the crown (in the build-up area), and into the cabinet top. Fasten the crown to the cabinet with glue and 2" wood screws. Plug the holes and cut the plugs off flush. Sand the plugs smooth and touch up with stain and varnish.

26 Lay the glass panels in the door rabbets. We used tempered glass, a slightly more expensive alternative to standard window glass that does not shatter into splinters when it breaks. Clamp the doors to the workbench. Set the glass stops in place and fasten with ¾" wire brads (**see Photo I**). Insert each brad at a 45° angle, being careful to avoid hitting the glass. A brad pusher is helpful for inserting the brads. Use a nailset to carefully drive the brad heads flush with the glass stops. Protect the glass around the work area with a piece of corrugated cardboard.

27 Drill guide holes then mount the knobs. Attach the hinges and hang the doors on the cabinet. If you wish, you can install an under-cabinet light behind the valance to illuminate a countertop area below the cabinet.

28 Level and attach the wall cleat to the wall with 2" screws, making sure to hit two wall studs with the screws. Mark the stud locations on the wall above. Hang the cabinet on the cleat, then drive 3" screws through the cabinet back and cabinet cleat and into the wall studs.

COUNTRY CUPBOARD

Add a touch of rural charm to any kitchen or dining area with this pine country cupboard. Our design features a beadboard back panel, sides made of glue-up pine panels (see the description on page 181), and arched accents to mimic traditional styling for storage cupboards of this kind. Both shelves are fully adjustable, and the cupboard is hung securely with concealed French cleats.

Vital Statistics: Country Cupboard

Type: Wall-hung country cupboard

Overall Size: 48" H x 36" W x 14" D

Material: Glue-up pine panels and plywood beadboard

Joinery: Dado, rabbet, and biscuit joints; butt joints reinforced with pocket screws

Construction Details:

- Decorative arch cutouts on valance, sides, and back panel
- Beadboard back panel fits into dadoes machined in cupboard sides
- Fully adjustable shelves
- Cupboard mounts to wall with interlocking French cleats

Finish: Two coats of orange or amber natural shellac

Building Time

 Preparing Stock: 3–4 hours

 Layout: 2–3 hours

 Cutting Parts: 3–5 hours

 Assembly: 2–3 hours

Finishing: 1–2 hours

Total: 11–17 hours

Tools You'll Use

- Compass or trammel points
- Table saw
- Jigsaw
- Band saw (optional)
- Drum or pad sander
- Biscuit joiner
- Drill/driver
- Pocket-hole jig
- Router table with ⅜" straight bit
- Hammer
- Clamps
- Chisel
- Combination square

Shopping List

- ○ ¾" x 24" x 48" glue-up pine panel
- ○ (2) ¾" x 16" x 6' glue-up pine panels
- ○ ¹¹⁄₃₂" x 4' x 8' pine beadboard
- ○ (20) brass shelf pin grommets
- ○ (8) ¼" brass shelf pins
- ○ #20 biscuits
- ○ 1¼" pocket screws
- ○ #8 flathead wood screws (1", 1½", 3")
- ○ Wood glue
- ○ Finishing materials

Exploded View

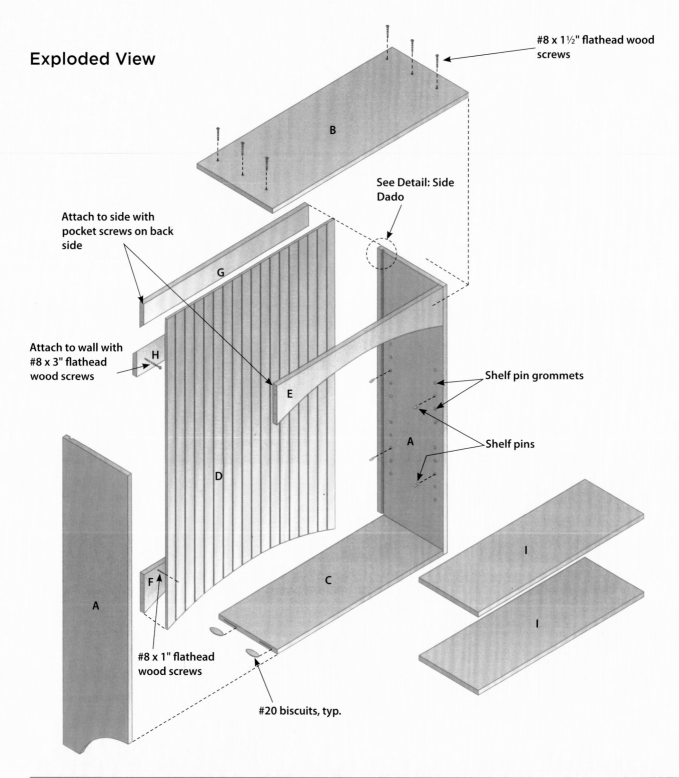

#8 x 1½" flathead wood screws

See Detail: Side Dado

Attach to side with pocket screws on back side

Attach to wall with #8 x 3" flathead wood screws

Shelf pin grommets

Shelf pins

#8 x 1" flathead wood screws

#20 biscuits, typ.

B

G

H

E

D

F

A

A

C

I

I

Wall-hung Cutting List

Part	No.	Size	Material	Part	No.	Size	Material
A. Sides	2	¾" x 12" x 47¼"	Glue-up panel (pine)	**F.** Blocking	1	¾" x 6" x 30½"	Glue-up panel (pine)
B. Top	1	¾" x 14" x 36"	"	**G.** Cabinet cleat	1	¾" x 4" x 30½"	"
C. Bottom	1	¾" x 10⅞" x 30½"	"	**H.** Wall cleat	1	¾" x 4" x 28½""	"
D. Back	1	1¹¹⁄₃₂" x 31" x 47⅝"	Pine beadboard	**I.** Shelves	2	¾" x 10 ¾" x 30½"	"
E. Valance	1	¾" x 6" x 30½"	Glue-up panel (pine)				

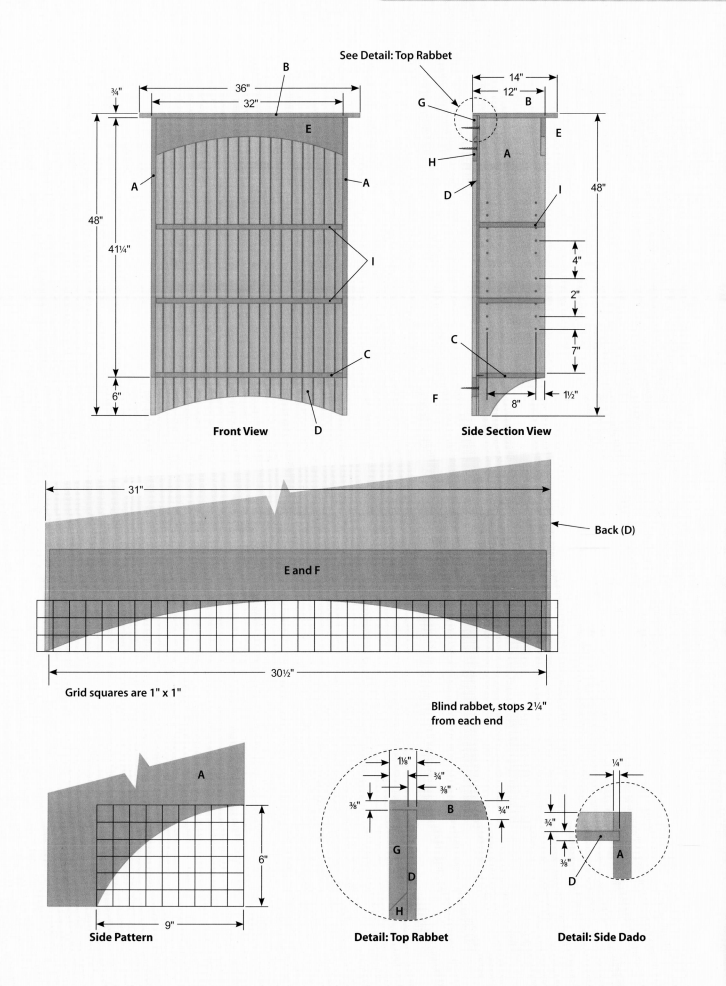

See Detail: Top Rabbet

Front View

Side Section View

Grid squares are 1" x 1"

Blind rabbet, stops 2¼"
from each end

Side Pattern

Back (D)

E and F

Detail: Top Rabbet

Detail: Side Dado

Country Cupboard: Step-by-Step

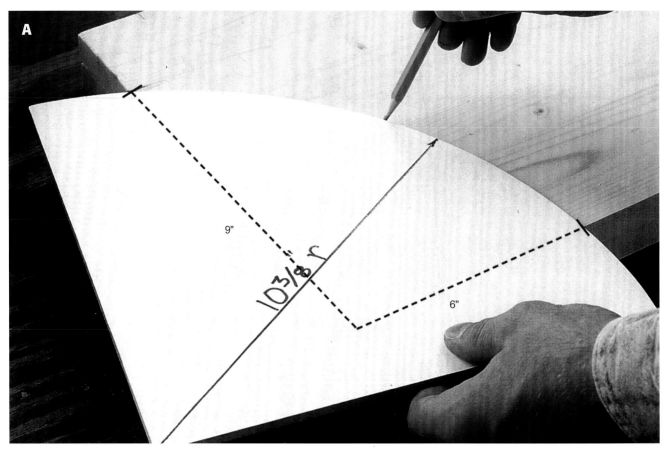

Photo A: Mark arched cutting lines on the side pieces using a template. The arc should intersect points measured from one corner—6" along one edge and 9" along the adjacent end of the panel.

Prepare the Sides and Shelves

1 Rip-cut and crosscut the sides and shelves to size from 12"-wide glued-up pine panels, according to the measurements given in the cutting list (see page 242). The sides feature a decorative arc cutout in the lower front corners. To lay out the arcs, make a reference mark 6" from one corner of one side panel along the edge and a second reference mark 9" from the same corner along the end of the panel. Make an arc template from ¼" scrap hardboard or plywood. Start with a square about 11" x 11". Use a compass or trammel points to swing a 10⅜"-radius arc from one corner. We attached white paper over the hardboard to see the pencil marks more clearly. Cut out the template and smooth the edges. Set the template on one side panel, aligning the arc with the layout marks. Then trace the arc (**see Photo A**).

2 Gang-cut the arcs on both side panels with a band saw or jigsaw. To do this, clamp both panels together so that the panel marked for the arc is on top and the edges and ends of the panels are flush. Cut along the layout line to form identical arcs. With the panels still clamped, sand both arcs smooth. Unclamp the panels.

3 Drill two rows of holes in each side panel to house the adjustable shelf pin grommets. Note: Grommets will keep metal shelf pins from widening the holes as they are inserted and removed. Draw two long lines that are parallel with the front and back edges of the sides and inset 1½" from each edge. Make a reference mark on both layout lines, 13¾" from the bottom edge of the sides, for the first holes. Then, mark off the rest of the holes as shown in the Side Section View, page 243. Drill ⁹⁄₃₂"-diameter holes at each mark, gauging the depth according to the grommets you buy. Use a depth stop on your drill bit.

4 Rout a ⅜"-wide dado on the inside face of each side panel for the back panel. Make each dado ¼" deep and locate it ¾" from the back edge of each panel. Cut the dadoes with a router and a ⅜" straight bit.

Photo B: Mark the arc cutout on the valance or bottom blocking by springing a thin, flexible wood or hardboard batten between nails to connect the centerpoint with the corners. The batten creates a smooth arc for tracing.

Photo C: We used spring clamps to stack the valance and bottom blocking so they could be gang-cut and sanded smooth together.

Make the Bottom, Valance, and Bottom Blocking

5 Cut the bottom, valance, and bottom blocking to size. The valance and bottom blocking have arc cutouts that are gang-cut and sanded simultaneously. Use a ¼"-thick strip of wood about 1" wide and about 42" long or a scrap piece of tempered hardboard to form a flexible batten for determining the arc curve. Measure and mark the center point (both lengthwise and widthwise) of either the bottom blocking or the valance and tack a long finish nail ¼" below this spot. Set the workpiece on a scrap board to use as a tacking surface for marking the ends of the arc, and nail a finish nail at each of these two locations. Flex the batten over the center point nail and under the nails at each end to form a gentle arc. Reinforce the batten's curve by adding a few more nails next to the batten. Draw a line along the batten to mark the curve (**see Photo B**).

6 Clamp the bottom blocking and valance together with the curve layout on top and gang-cut the curves in both workpieces. Smooth the curves with a sander (**see Photo C**).

7 Drill three pocket screw holes in each end of the valance and bottom blocking in order to attach these pieces to the side panels later. (For more on pocket screws, see page 185.) Drill the holes on the back side of each piece so the screws won't show once the cupboard is assembled.

8 Lay out and cut #20 biscuit slots for attaching the ends of the bottom panel to the sides. Use one end of the bottom as a straightedge to help align the biscuit joiner as you cut slots in the side panels. Glue biscuits into the ends of the bottom now to make assembly easier later, and wipe up excess glue around the biscuits.

Prepare the Back
and Top Panels

9 Cut the back panel to size from
11/32" pine beadboard. Using the
bottom blocking or valance as a pattern,
trace the decorative arc onto the bottom of
the back. Cut the back with a jigsaw and
sand the edge smooth.

Note

If your router fence splits into
two pieces as ours does, simply
set the fence halves slightly over
the edge of the bit instead of
cutting into a scrap-wood fence.

10 Cut the top to size. Rout a ⅜-deep
rabbet along the back, bottom
edge of the top. This stopped (also called
blind) rabbet should stop 2" from either end.
Mark the workpiece to indicate starting and
stopping points for the rabbet. Also mark
the fence to show the left (outfeed) and
right (infeed) edges of the bit. Then install a
½"- or ¾" straight bit for cutting the rabbet.
To prepare for the first cut, attach a scrap
wood fence to your router table fence. Set
the scrap fence as close as possible to the
blade, without touching it. Raise the blade
to ⅜" cutting depth. Start the router and
slide the router fence forward until the bit
cuts into the scrap fence about ⅛". Begin the
rabbet by starting the router with the
workpiece clear of the bit. Pivot the
workpiece slowly into the bit and against
the fence so the starting mark on the
workpiece aligns with the outfeed blade
mark on the fence. Then feed the workpiece
along the fence until the stop mark on the
workpiece aligns with the infeed blade mark
(**see Photo D**). Since the rabbet is 1⅛" wide,
you'll need to make multiple side-by-side
passes in this fashion to cut the rabbet to full
width, sliding the fence ¼" or so back from
the bit with each pass, removing
progressively more material. Once you are
finished routing, square up the ends of the
rabbet with a chisel.

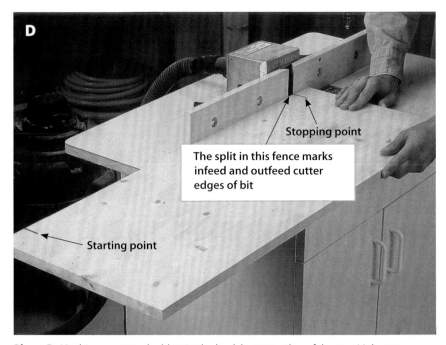

Stopping point

The split in this fence marks
infeed and outfeed cutter
edges of bit

Starting point

Photo D: Machine a stopped rabbet in the back bottom edge of the top. Make stop
marks on the workpiece to locate where to start and end the cut, and make a second set
of marks to locate the edges of the bit. Use the workpiece and bit marks in tandem to cut
the rabbet.

Photo E: Dry-assemble the carcase with clamps and pocket screws to check the fit. Start
by installing the valance and bottom panel temporarily with screws and clamps.

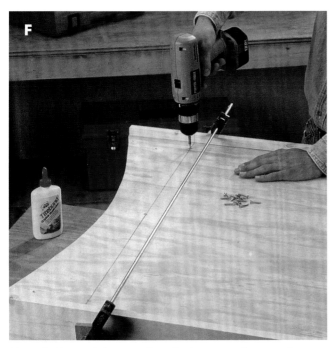

Photo F: Glue and clamp up the sides, bottom, valance, and top. Slide the back in place and screw it to the cupboard bottom with 1" flathead wood screws. Check the cupboard for square by measuring the diagonals before driving the screws.

Photo G: Attach the cabinet hanging cleat and bottom blocking to the cupboard sides with glue and pocket screws. Be sure the cabinet-hanging cleat bevel faces the right direction (see page 193).

Assemble the Cupboard

11 Dry-assemble the cupboard. Set and clamp the valance and bottom pieces into place between the side panels, with the front of the cupboard facing down on a suitable worksurface. Drive pocket screws into the valance pocket holes, stopping short of seating the screw heads (**see Photo E**). Mark the centerline of the cupboard bottom on the back inside edge of each side panel, behind the dado. Slide the back into the dado. Set the top into position over the side panels. Screw the top temporarily to the sides using 1½" flathead wood screws. Slide the back in the dado; the bottom of the back should sit flush with the bottom of the sides. Draw a line across the back to connect the bottom panel centerline marks.

12 Disassemble the cupboard and apply glue to the ends of the bottom, valance, and top edges of the sides. Reassemble and clamp up the cupboard carcase, wiping up any glue that squeezes out with a damp cloth. Slide the back into the dado and attach it to the bottom with 1" flathead wood screws (**see Photo F**). Attach the bottom blocking piece to the cupboard sides with glue and pocket screws, aligning the bottom blocking and back panel arcs.

13 Cut the hanging French cleats to size and rip a 45° bevel along one long edge of each cleat. Attach the cabinet hanging cleat to the cupboard sides with glue and pocket screws, orienting the cleat as shown in Detail: Top Rabbet, page 243 (**see Photo G**).

Finishing Touches

14 Plug the screw holes in the top if you like, and trim the plugs flush when the glue dries. Sand the cupboard with 120-grit sandpaper and apply the finish of your choice. Traditionally, country pine furnishings are finished with natural shellac, so we applied two coats of amber shellac flakes dissolved in alcohol.

15 Push metal grommets into the shelf pin holes in the sides. If it's difficult to drive them in fully, tap the grommets home with a hammer and block of wood.

16 Attach the wall hanging cleat to the wall with 3" countersunk flathead wood screws. Since the wall cleat is 2" shorter than the cabinet cleat, you have some leeway from side to side when hanging the cupboard. Make sure you hit two wall studs when you drive the screws to anchor the cleat solidly. Mark stud locations on the wall so you can drive screws into the studs from inside the cupboard after hanging it on the wall cleat.

17 Hang the cabinet on the wall and drive 3" screws into the wall studs at the top and bottom of the cabinet (through the cabinet hanging cleat and bottom blocking). Install the shelf pins and adjustable shelves.

HALLWAY BOOKCASE

Straightforward to build and a practical solution for adding hallway storage space, this contemporary bookcase may be just the impetus you need for organizing those photo albums, encyclopedias and book collections. Our painted MDF bookcase provides nearly 3' of display space per shelf, but its angled front profile and slender 10" depth allow it to nestle conveniently along the wall of even a narrow hallway. The shelves tip back slightly to keep items from toppling off if they should get brushed by passersby. We've customized the bookcase design with knockdown hardware so it can be easily disassembled for storage.

Vital Statistics: Hallway Bookcase

Type: Bookcase

Overall Size: 60" H x 36" W x 10" D

Material: MDF, solid poplar

Joinery: Rabbet and butt joints

Construction Details:

- Unit is easily assembled and disassembled with "knockdown" mechanical fasteners

- MDF panels are paint-ready and require no edge banding

- Solid-wood shelf stiffeners prevent the shelves from sagging

- Angled front profile adds a contemporary look

Finish: Primer and enamel paint

Building Time

 Preparing Stock: 1 hour

 Layout: 1–2 hours

 Cutting Parts: 2–3 hours

 Assembly: 2–3 hours

 Finishing: 1–2 hours

Total: 7–11 hours

Tools You'll Use

- Table saw
- Circular saw with straightedge guide
- Combination square
- Drill press
- Router
- ½"-rad. piloted roundover bit
- ½" piloted rabbeting bit or straight bit and router edge guide
- Dado-blade set (optional)
- Biscuit joiner (optional)
- Drill/driver
- Pocket-screw jig

Shopping List

- ¾" x 4' x 8' MDF
- ½" x 35½" x 5' MDF
- ¾" x 1½" x 16' poplar
- (20) mechanical fasteners
- #8 × 1¼" flathead wood screws
- #20 biscuits (optional)
- 1¼" pocket screws
- Wood glue
- Finishing materials

Exploded View

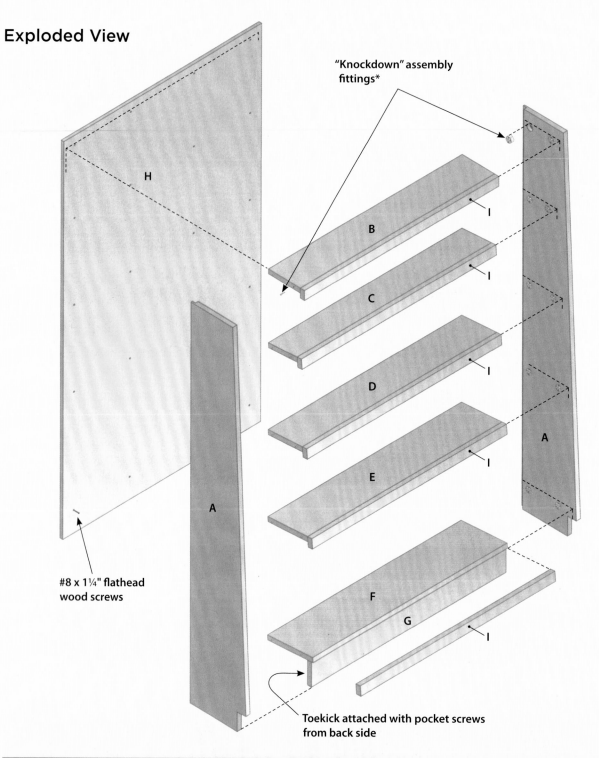

"Knockdown" assembly fittings*

I

H

B

C

I

D

I

E

I

A

A

#8 x 1¼" flathead wood screws

F

G

I

Toekick attached with pocket screws from back side

Wall-hung Cutting List

Part	No.	Size	Material	Part	No.	Size	Material
A. Sides	2	¾" x 10" x 60"	MDF	**F.** Bottom	1	¾" x 8¼" x 34½"	MDF
B. Top	1	¾" x 5¾" x 34½"	"	**G.** Toekick	1	¾" x 3¾" x 34½"	"
C. Shelf	1	¾" x 5¼" x 34½"	"	**H.** Back	1	½" x 35½" x 59¾"	"
D. Shelf	1	¾" x 6¼" x 34½"	"	**I.** Shelf stiffeners	5	¾" x 1½" x 34½"	Poplar
E. Shelf	1	¾" x 7¼" x 34½"	"				

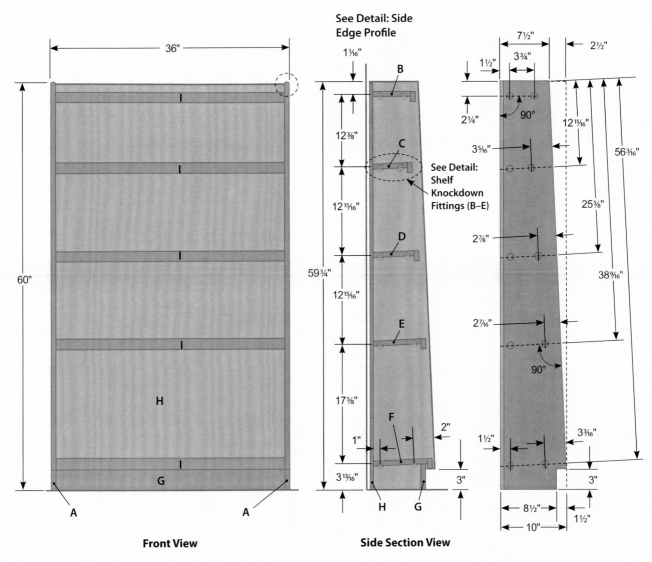

See Detail: Side Edge Profile

1¹⁄₁₆"

B

12³⁄₈"

C

See Detail: Shelf Knockdown Fittings (B–E)

12¹⁵⁄₁₆"

D

59¾"

12¹⁵⁄₁₆"

E

17⅝"

F

2"

1"

3¹³⁄₁₆"

3"

H   G

Front View

36"

60"

H

I

G

A   A

Side Section View

7½"

3¾"

2½"

1½"

90°

2¼"

3⁵⁄₁₆"

12¹¹⁄₁₆"

56³⁄₁₆"

2⅞"

25⅝"

2⁷⁄₁₆"

38⁹⁄₁₆"

90°

1½"

3³⁄₁₆"

8½"

3"

10"

1½"

Drilling Pattern (Side)

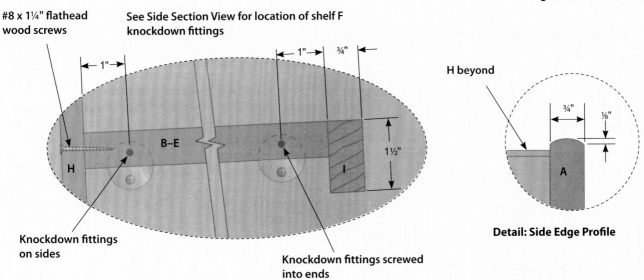

#8 x 1¼" flathead wood screws

See Side Section View for location of shelf F knockdown fittings

1"   1"   ¾"

B–E

1½"

H   I

Knockdown fittings on sides

Knockdown fittings screwed into ends

Detail: Shelf Knockdown Fittings Drilling Pattern (B–E)

H beyond

¾"   ⅛"

A

Detail: Side Edge Profile

Hallway Bookcase: Step-by-Step

Construct the Sides

1 Cut the sides to size from ¾" MDF. A carbide-tipped plywood or panel-cutting blade works best for cutting MDF. Use the Drilling Pattern (Side) drawing, page 251, as a guide for laying out the angled profile along the front edge of each side. The top edge of each side is 7½" wide. Use a straightedge to draw a line connecting this point with the lower front corner on one side piece. Now clamp the two sides together with the marked side facing out. Gang-cut the angled front profiles using a circular saw and a straightedge guide. Unclamp the side panels.

2 Lay out the toekick notches in the bottom front corner of each side using the measurements in the drawing. The notch is 3" high and 1½" deep.

3 Mark locations for the barrel-shaped knockdown housings that will be inset into the sides to support the top, shelves, and bottom. The top is perpendicular to the back edge of the sides, so measure 2¼" down along the back edge and draw a line across the width of each side panel with a combination square. Use the Drilling Pattern (Side) drawing, page 251, as a guide for laying out knockdown hardware positions for the shelves and bottom. Measure down from the top front corner of each side and make a reference mark for each shelf and the bottom. The three shelves and bottom are perpendicular to the front edge, so use a combination square registered against this edge to extend the centerlines across the width of each side (**see Photo A**).

4 Locate center points for drilling holes for the knockdown housings along the layout lines you drew in step 3. The rear fittings are 1½" in from the back edge of each side for the top and shelves and 1" for the bottom. To establish the front hole for the top, measure over 3¾" from the rear hole center mark. Measure from the front edge of the sides to determine front hole drilling locations for the shelves and bottom. The shelf hole placement varies.

Photo A: Mark centerlines for locating the mechanical fasteners in the side panels. Use a combination square to draw lines perpendicular to the front edges of the sides for the shelves and bottom.

Tip

If you don't own a drill press, you could also drill these holes with a right-angle drilling guide and a portable drill. Be sure to mark the bit depth with a strip of masking tape so you don't inadvertently drill too far.

5 Drill the knockdown housing holes on a drill press. The fasteners I used require a 25-mm Forstner bit, set to drill ½"-deep holes. Test the drilling setup on a piece of ¾" scrap first. Drill all the holes in both sides (**see Photo B**). Clamp each side panel to the drill press when drilling these holes.

6 Cut out the toekick notch at the base of each side with a jigsaw and a fine-tooth blade to avoid chipping the cut edges. Sand the edges smooth.

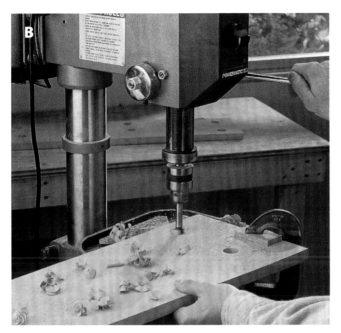

Photo B: Drill 25-mm holes for the mechanical fasteners, ½" deep into the side panels, using a Forstner bit in the drill press. Clamp each side panel to the drill press table to steady the parts as you drill.

Photo C: Ease the top and front edges of the side panels using a router with a piloted ½" roundover bit. Center the bit's bearing across the thickness of the panel to form a ⅛" "thumbnail" roundover.

7 Ease the top and front edges of the sides. To accomplish this, install a ½" piloted roundover bit in your router. Set the bit depth so the bearing rides along the stock about halfway across its thickness. This bit setting should produce a roughly ⅛"-radius "thumbnail" roundover profile. Test the setup on a piece of ¾" scrap MDF to determine the appropriate bit setting. Rout this profile along the top and front edges of both sides (**see Photo C**). Smooth the profiles with sandpaper.

8 Cut rabbets along the inside back edges of the sides to house the bookcase back panel. Cut these rabbets with a piloted ½" rabbeting bit or a straight bit and edge guide. You could also cut these rabbets on the table saw with a dado-blade set. The rabbets extend the full length of the sides.

Mechanical Fasteners

The top, bottom, and shelves of this bookcase are attached to the sides with knockdown (KD) mechanical fasteners, often called *RTA (ready-to-assemble) fittings*. The hardware consists of a barrel-shaped plastic housing with a cam-style screw that fits into a 25-mm hole drilled into the vertical member of the joint. A coarse-threaded pin, fastened into the edge or end of the horizontal joint member, slides into a hole above the cam screw in the knockdown housing. By turning the cam screw clockwise, it intersects with the shelf pin, locking the joint parts together. To unlock the joint, simply turn

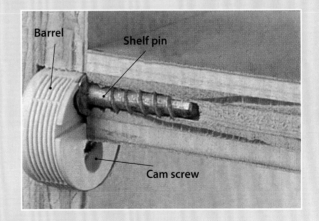

the cam screw counterclockwise to release the pin. Knockdown hardware is a good solution for furniture that may need to be disassembled from time to time. Driving and removing ordinary screws would quickly strip screw threads in the wood, especially in materials like MDF or particleboard.

Prepare the Top, Bottom, and Shelves

9 Cut the top, bottom, and shelves to size. Drill holes for the knockdown connector pins in the ends of these parts. One way to ensure that your holes are straight and square is to drill these holes on a drill press with the table turned vertically and the part clamped flat against it (**see Photo D**). Tilt the drill press table and adjust it for plumb with a level. Use a ³⁄₁₆" drill bit and a depth stop to drill two 1⅛"-deep holes in the end of each part, centered on the thickness of the boards. Locate the rear holes 1" from the back edge of each workpiece. On all parts but the bottom, the front holes should also be placed 1" from the front edges. Drill the front holes in the bottom 2" from the front edge.

10 Rip-cut and crosscut five shelf stiffeners from ¾"-thick poplar. Clamp the stiffeners to the top, bottom, and shelf panels, making sure the top edge of each stiffener is flush with the top face of each shelf and the ends of the stiffeners line up with the shelf ends. Fasten the stiffeners in place with glue and 6d finish nails (**see Photo E**). Drill pilot holes for the nails to keep from splitting the poplar and the MDF. (Option: You could also cut #20 biscuit joints before installing the stiffeners to help align the parts.) Clean up excess glue with a chisel or scraper before it hardens.

Cut the Remaining Carcase Parts and Dry-Assemble the Bookcase

11 Cut the back panel to size from ½" MDF, and cut the toekick from ¾" MDF. Ease the top edges of the back using the same router technique you used for rounding over the top and front edges of the bookcase sides. You'll have to adjust the bit depth so the bearing rides along the center of the back panel thickness. Test on ½"-thick scrap first. Smooth the roundovers with sandpaper.

12 Dry-assemble the bookcase to check the fit of the parts. Be sure the top, bottom, shelves, and toekick fit snugly between the side panels with the back panel set into place in the side-panel rabbets. Then disassemble the bookcase.

13 Prime and paint the faces, edges, and ends of all the parts. You can prime and paint at this stage because all the fasteners used to assemble the bookcase will be hidden from view, so there is no need for covering nail or screw holes before finishing. Finishing the parts now is also an advantage because you have full access to all part surfaces.

Photo D: Bore 1⅛"-deep holes in the ends of the top, bottom, and shelf parts for the knockdown connector pins. Drill two ³⁄₁₆"-diameter holes in each end. We swiveled our drill press table vertically and clamped the parts to it to ensure square, straight holes.

Photo E: Attach the shelf stiffeners to the front edges of the top, bottom, and shelves. Clamp the stiffeners into place, aligning the top edge of the stiffeners with the top face of the MDF panels. Fasten the stiffeners with glue and 6d finish nails. For easier alignment, you could also cut #20 biscuit slots between the parts.

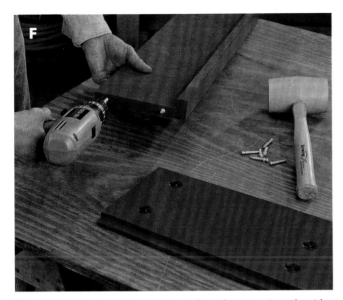

Photo F: Tap the barrel-shaped knockdown housings into the side panels with a rubber mallet. Then screw the pin connectors into the ends of the top, bottom, and shelves. Do not overdrive the pins.

Photo G: Assemble the bookcase by inserting the top, bottom, and shelves into one side panel and tightening the cam screws. Lay the assembly on its face and install the other side. Attach the toekick.

Install the Mechanical Fasteners

14 Press the barrel knockdown housings into the 25-mm holes in the side panels. Orient the straight groove in each plastic housing so it runs perpendicular to the front edge of the side panels, with the cam screw facing down toward the bottom of the bookcase. This way, you'll be able to set the top, bottom, and shelves into place in the bookcase and tighten the knockdown fittings from just underneath the shelves. Tap the housings to seat them fully into their holes so the barrels sit flush with the inside faces of the sides.

15 Insert the knockdown connector pins into the ends of the top, bottom, and shelves (**see Photo F**). Use a #2 Phillips screwdriver or a Phillips bit in a drill/driver to twist the connector pins into the holes until the collars on the pins seat against the workpieces. If your drill/driver has an adjustable clutch, dial it to a low setting to keep from overdriving the pins and stripping the holes in the parts.

Assemble the Bookcase

16 Attach the top, fixed shelves, and bottom to one side panel. Be sure to install the parts so the shelf stiffeners face the front edge of the bookcase. When assembling these parts, insert the pin connectors into the knockdown housings and tighten the joints by turning the cam screws clockwise (**see Photo G**). Lay the assembly on its face and attach the other side panel. Clamp the toekick in place and fasten it with pocket screws, drilling from the back side of the toekick into the side panels. Drive one pocket screw through the toekick and into the bottom of the bottom panel as well.

Photo H: Drill pilot holes through the back panel and into the top, bottom, and shelves, and attach the back to the bookcase with 1¼" flathead wood screws.

17 Measure and mark centerlines for the top, bottom, and shelves onto the back edges of the sides to serve as reference marks for fastening the back panel. Set the back into the side panel rabbets and align the bottom edges of the back and sides. Draw screw reference lines across the back panel, connecting the marks you drew on the back edges of the sides. Drill countersunk pilot holes through the back and into the horizontal members of the bookcase with 1¼" flathead wood screws (**see Photo H**).

WALNUT WRITING DESK

Create a dedicated space for organizing your busy life when you build this Shaker-inspired writing desk. It's an attractive furnishing for the library, den, or any other room in your home. The table's clean, understated styling and rich walnut wood tones give it the look and feel of solid hardwood construction—but here's the secret: Our table is built almost entirely of veneered plywood and MDF at a considerable savings over building with solid walnut. The table's legs taper on two faces and can be removed from the table, should the table need to be moved or stored.

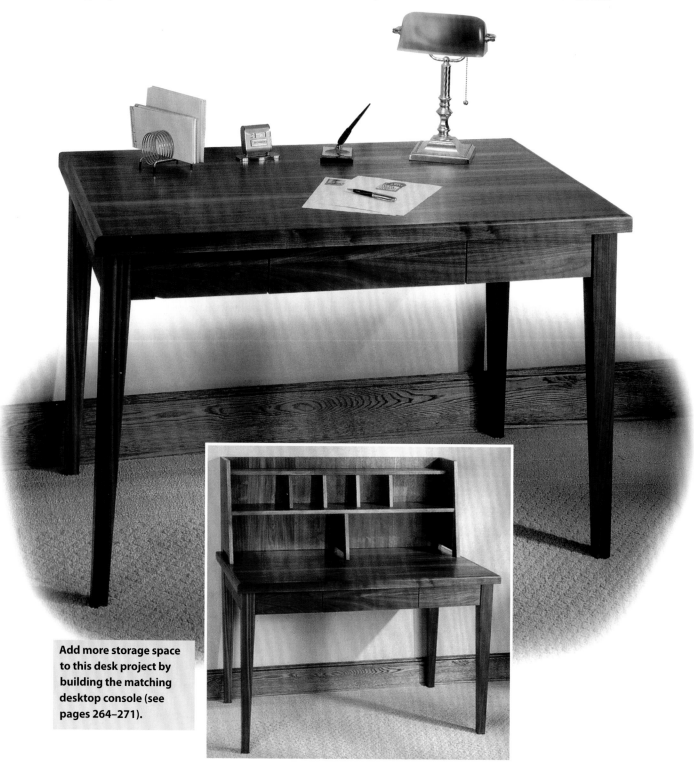

Add more storage space to this desk project by building the matching desktop console (see pages 264–271).

Vital Statistics: Walnut Writing Desk

Type: Writing desk with drawer

Overall Size: 30" W x 48" L x 30" H

Material: Walnut-veneer plywood, walnut veneer, solid walnut, birch plywood, MDF

Joinery: Biscuit joints, face-glued laminated joints, tongue-and-groove corner joints, miter joints

Construction Details:

- Legs are face-glued MDF, tapered, and veneered
- Metal corner brackets attach legs to aprons
- Walnut plywood is cut so that grain matches across front aprons and drawer front
- Iron-on veneer edging tape
- Plywood drawer fitted with flush-fitting drawer front; tongue-and-groove construction

Finish: Two coats of satin polyurethane varnish

Building Time

 Preparing Stock: 2 hours

 Layout: 2–4 hours

 Cutting Parts: 2–4 hours

 Assembly: 2–3 hours

 Finishing: 2–3 hours

Total: 10–16 hours

Tools You'll Use

- Table saw
- Tapering jig
- Clamps
- Band saw
- Jointer
- Biscuit joiner
- Drill/driver
- Right-angle drilling guide or drill press
- Pocket-screw jig
- Router with ⅜" roundover bit

Shopping List

- ○ ¾" x 48" x 48" walnut plywood
- ○ ¾" x 48" x 48" MDF
- ○ ½" x 18" x 48" birch plywood
- ○ ¼" x 24" x 24" birch plywood
- ○ ¹⁄₆₄" x 10" x 48" walnut veneer
- ○ (4) ¾" x 1½" x 4' walnut
- ○ ¹³⁄₁₆" x 8' walnut veneer edge banding
- ○ (4) metal leg corner brackets
- ○ (8) ¼" x 2" hanger bolts, washers, and wing nuts
- ○ (2) 20" full-extension drawer slides
- ○ (8) 1" brass L-braces
- ○ #20 biscuits
- ○ #8 x 1¼" flathead wood screws

Exploded View

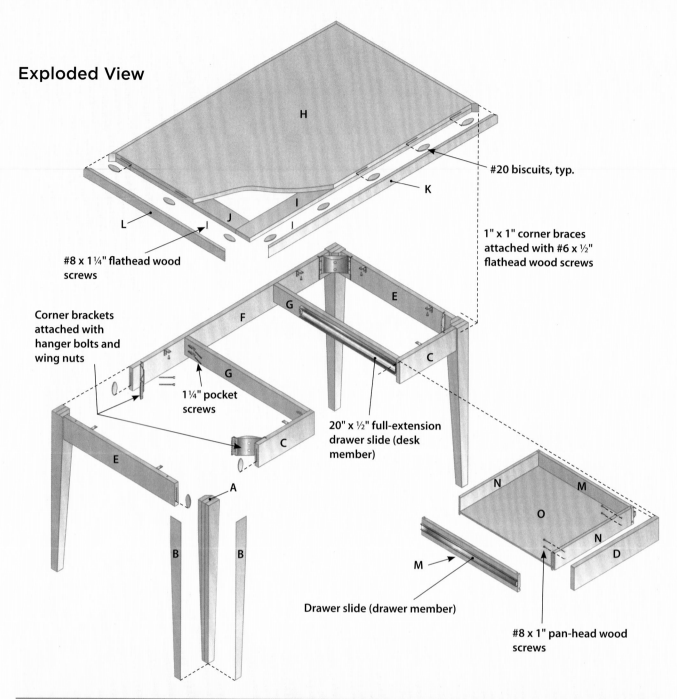

#20 biscuits, typ.

1" x 1" corner braces attached with #6 x ½" flathead wood screws

#8 x 1¼" flathead wood screws

Corner brackets attached with hanger bolts and wing nuts

1¼" pocket screws

20" x ½" full-extension drawer slide (desk member)

Drawer slide (drawer member)

#8 x 1" pan-head wood screws

Writing Desk Cutting List

Part	No.	Size	Material	Part	No.	Size	Material
A. Leg blanks	12	¾" x 2½" x 29"	MDF	**I.** Front/back build-up	2	¾" x 3" x 46½"	MDF
B. Leg veneer	16	¹⁄₆₄" x 2½" x 29"	Walnut veneer	**J.** Side build-up	2	¾" x 3" x 22½"	"
C. Front apron	2	¾" x 4" x 11¾"	Walnut plywood	**K.** Front/back edge	2	¾" x 1½" x 48"	Solid walnut
D. Drawer front	1	¾" x 3⅞" x 17 ¹¹⁄₁₆"	"	**L.** Side edge	2	¾" x 1½" x 30"	"
E. Side apron	2	¾" x 4" x 23½"	"	**M.** Drawer side	2	½" x 3" x 20"	Birch plywood
F. Back apron	1	¾" x 4" x 41½"	"	**N.** Drawer front/back	2	½" x 3" x 16½"	"
G. Spreader	2	¾" x 3" x 26"	"	**O.** Drawer bottom	1	¼" x 19½" x 16½"	"
H. Tabletop	1	¾" x 28½" x 46½"	"				

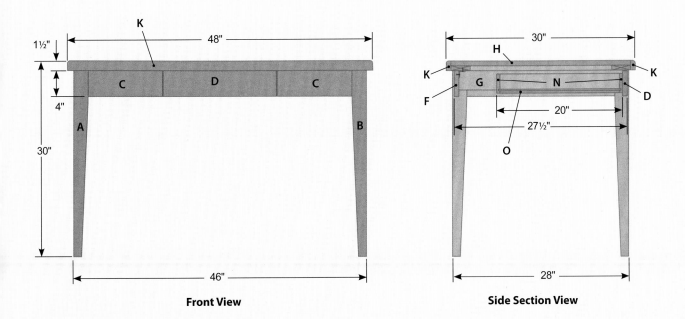

Front View

Side Section View

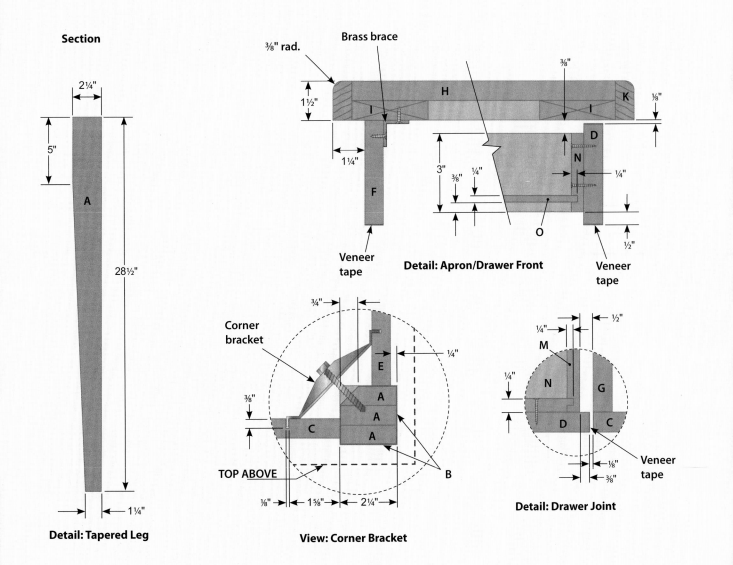

Section

Detail: Tapered Leg

3/8" rad.

Brass brace

Detail: Apron/Drawer Front

Veneer tape

Veneer tape

Corner bracket

TOP ABOVE

View: Corner Bracket

M

Veneer tape

Detail: Drawer Joint

Walnut Writing Desk: Step-by-Step

Make the Legs

1 Cut 12 leg blanks to size from ¾" MDF. The blanks are oversized so the legs can be cut to size after they are face-glued into four separate three-board assemblies. Spread glue on the faces of the blanks and clamp up a stack of three blanks for each leg. Use plenty of evenly spaced clamps spaced about 6" apart to hold the glue joints closed. After the glue dries, scrape off glue squeeze-out with a paint scraper (wear safety glasses to protect your eyes from flying glue chips).

2 Rip-cut the legs to a finished width of 2¼" and crosscut each leg to 28½".

3 Taper two adjacent faces of each leg. The taper starts 5" down from the top of the leg and extends to the bottom of the leg, reducing it to 1¼" square. Draw layout lines to mark the tapers and use a tapering jig on the table saw to cut the tapers. Adjust the jig's angle and the saw's rip fence so the blade will follow the marked taper line as the jig and leg are pushed together along the rip fence. Cut the tapers, holding each leg tightly against the tapering jig with a push stick (**see Photo A**). Once you've cut the first taper, turn the leg in the jig so the taper side faces up and taper a second leg face. Sand away any surface unevenness left by the saw blade.

Photo A: Cut tapers on two adjacent sides of each leg using a table saw and tapering jig. When cutting MDF, wear a particle mask—sawing MDF generates fine sawdust.

> ### Tip
> Line the cauls with wax paper to keep any glue that migrates through the veneer from bonding the cauls to the veneer.

4 Cover the leg faces with walnut veneer. Use a veneer saw or a utility knife to cut strips of walnut veneer for the four faces of each leg. When cutting the veneer, guide your veneer saw or knife blade against a straightedge, especially when slicing in the direction of the grain, as blades have a tendency to follow and split the grain. Cut the veneer strips about ¼" oversize in width and length to overhang the leg faces. Spread a thin layer of glue evenly over two opposite leg faces (not on the veneer), and clamp the veneer in place, using scrap-wood cauls between the clamp jaws and the veneer. The cauls should cover the veneer on the leg faces completely to press the entire sheets of veneer flat. The tapered faces of the legs will need two separate cauls—one for the leg portion above the taper and the other to cover the taper. After the glue dries, trim off excess veneer carefully with a router and flush-trimming bit, a block plane, or a sharp chisel. Try to cut with the grain rather than against it, so any grain that splits will fracture away from, rather than into, the veneer covering the leg face. Then veneer the remaining two faces of each leg in the same way.

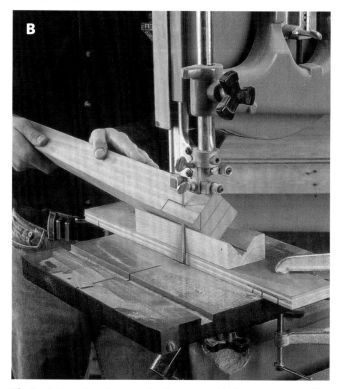

Photo B: Use a band saw and a shop-made V-block support cradle to cut a stopped chamfer on the top inside corner of each leg. Make the long chamfer cuts first, then swivel the jig 90° to cut the waste pieces free, creating recesses for the leg hardware.

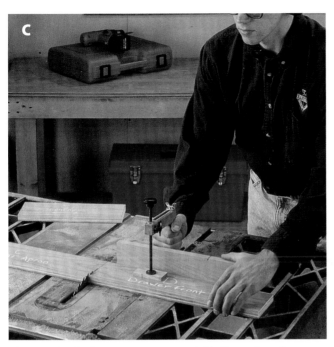

Photo C: Cut the front apron pieces and the drawer front from one piece of walnut plywood so the grain will match on all three parts.

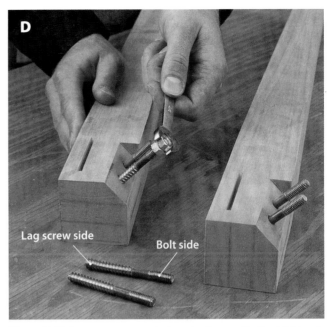

Lag screw side

Bolt side

Photo D: Tighten two nuts against each other on the "bolt" side of the hanger bolts to serve as a "head." Use a wrench to twist the "lag screw" side of each hanger into its hole in the legs. Then remove the nuts.

5 Cut a stopped chamfer along the top inside corner of each leg (the corner that separates the tapered faces of the legs). The chamfers provide clearance for metal corner braces that will be used to attach the legs and table aprons. To lay out the chamfers, draw a pair of 3"-long layout lines ¾" from, and parallel to, the inside leg corner along the adjacent leg faces. Cut the chamfers on a band saw using a "V-block" cradle to hold the leg at a 45° angle to the blade (**see Photo B**). To make the cradle jig, joint the edges and faces of a 2 x 4 straight and square. On the table saw, make two passes with the blade set to 45° to cut a 1"-deep, V-shaped notch along the face of the 2 x 4. Make your bevel cuts so the V-notch is positioned next to a long edge. Attach the cradle to a piece of plywood long enough to clamp to the saw table and flush with one edge of the plywood. Set a leg in the V-notch so the chamfer layout extends beyond the edge of the plywood and lines up with the saw blade. Clamp the jig in place on the saw table. Slide all four legs along the cradle, cutting the chamfer up to the chamfer stop line. Then unclamp the cradle setup, swivel it perpendicular to the blade, and use it to support the legs in order to crosscut the chamfer waste pieces free.

Assemble the Legs and Apron

6 Cut the front apron, drawer front, side aprons, back apron, and stretchers to size from ¾" walnut plywood. Cut the front apron pieces and the drawer front from one long, 4"-wide strip of plywood so the grain on the parts will match all the way across the front of the table (**see Photo C**). Trim ⅛" off the top long edge of the drawer front to allow for edge-banding and clearance space under the tabletop.

7 Cut #20 biscuit slots into both ends of the back and side aprons and the ends of the front aprons opposite the drawer front. Center the slots on the apron ends. Inserting biscuits at these locations will help align the aprons with the legs during assembly. Then cut a #20 biscuit slot into both chamfered faces of each leg but not in the chamfered areas. Position these slots ⁹⁄₁₆" from the edge opposite the chamfer, and center the slots 2" below the top of the leg. This way, you'll create a ¼" decorative reveal between the legs and aprons. Glue biscuits into the slots in the ends of the aprons and clean off excess glue.

8 Fit the aprons and legs together upside down on the workbench and set the corner brackets in place in the leg chamfers. Mark bracket hanger bolt hole locations on the legs in the chamfered areas. Also mark the insides of the aprons where grooves will need to be cut to receive the metal "lips" on either end of each bracket. Disassemble the apron. Drill straight holes for the hanger bolts into each leg using a right-angle drilling guide or by clamping the V-block to the drill press table. Install the hanger bolts by threading two nuts onto the bolt and tightening the nuts against one another. Then grip the top nut with a wrench and screw the bolt into the wood (**see Photo D**). Remove the nuts.

9 Cut ¼"-deep grooves following your bracket layout lines to house the bracket lips. Apply iron-on walnut veneer edge tape to the bottom edges of the back, side, and front aprons. Edge-band the ends of the front aprons that will face the drawer, as well as all four edges of the drawer front. For more information on applying edge banding, see page 186.

Photo E: Assemble the legs and apron parts by attaching the legs to the corner brackets and aprons with wing nuts. Tighten the wing nuts to close the joints—no glue is required.

Photo F: Attach the tabletop to the apron assembly with brass L-braces and ½" screws. Adjust the brackets so the screws pull the tabletop snugly against the apron assembly.

Tip

It isn't necessary to glue the apron biscuits into the leg slots for strength. The corner brackets will hold the legs securely and make leg removal easy in the future, should you need to disassemble the table for transport.

10 Reassemble the apron parts on the workbench and square up the assembly by measuring diagonally between the legs. Use wing nuts to fasten the legs to the corner brackets and aprons (**see Photo E**).

11 Cut the spreaders to size. Drill holes for pocket screws on the ends of the spreaders and screw them to the front and back aprons. Position each spreader 11" from the closer leg pair, and be sure that the top edges of the spreaders are flush with the tops of the aprons.

Construct and Attach the Tabletop

12 Cut the tabletop to size from ¾" walnut plywood. Cut the front, back, and side build-ups from ¾" MDF. Attach the build-ups to the underside of the tabletop with glue and countersunk 1¼" flathead wood screws, keeping the build-up edges flush with the outside edges of the tabletop.

13 Cut the front, back, and side tabletop edging from ¾" solid walnut, leaving the ends long. Miter-cut the ends of the edging so the pieces fit snugly around the tabletop. Cut #20 biscuit slots about every 12" to aid in aligning the edging. Spread glue on the mating surfaces and the mitered corners and clamp the edging in place. Use pads between the clamp jaws to protect the edging. After the glue dries, remove the clamps and rout a roundover around the top edge of the tabletop using a ⅜"-diameter roundover bit.

14 Center the apron assembly on the bottom of the tabletop and attach the aprons to the tabletop build-ups with brass L-braces. Screw two braces to each apron, fastening them in place with ½" screws (**see Photo F**). Position the braces on the front aprons near the spreaders to anchor the spreaders and aprons where they bear the weight of the drawer. Since the metal L-braces are slotted to allow for adjustment, position them slightly below the top edge of each apron. This way, when you fasten each bracket to the tabletop build-ups, the screws will pull the tabletop tight against the apron.

Photo G: Glue and clamp up the drawer parts with the drawer bottom captured in its groove. Use short wood cauls to apply clamping pressure along the whole wood joint.

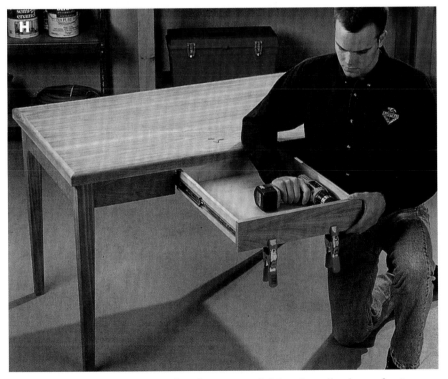

Photo H: With the drawer mounted in place on metal slides, clamp the drawer front temporarily in position with spring clamps. Attach the drawer front to the drawer box with screws.

Build the Drawer

15 Cut the drawer front, back, and sides to size from ½" birch plywood. Cut the drawer bottom from ¼" birch plywood. Use a table saw and dado-blade set or a router table and straight bit to construct interlocking ¼" x ¼" rabbet joints on the ends and edges of the drawer parts (see Detail: Drawer Joints, page 259, to configure the joints). Cut a ¼" x ¼" groove along the inside faces of all four drawer parts, ⅜" up from the bottom edges, to capture the drawer bottom. Finish-sand the interior faces of all drawer parts. Apply glue to the joint rabbets (use no glue in the drawer bottom groove), slide the drawer bottom into place, and secure the drawer box parts with clamps and cauls (**see Photo G**). Wipe away any excess glue with a damp cloth. Check the drawer for squareness by measuring across the diagonals, and reposition the clamps as needed to correct for square.

Finishing Touches

16 Finish-sand all table parts and apply the finish of your choice. We used two coats of satin polyurethane varnish, rubbing between each dry coat with 0000-fine steel wool. If you like, you can stain the birch drawer parts to match the rest of the walnut before applying the varnish.

17 Attach metal drawer slide hardware to the table spreaders and drawer box according to the manufacturer's instructions. Center the drawer in the opening so the bottom edges of the front aprons and the drawer face align.

18 Slide the drawer into the apron and attach the drawer front to the drawer box. Adjust the position evenly beneath the tabletop and the front aprons using small spring clamps to temporarily hold the drawer front in place from below. Once you are satisfied with the position of the drawer front, drill pilot holes and install 1" pan-head screws from inside the drawer box (**see Photo H**).

WALNUT DESKTOP CONSOLE

Add a new dimension to a desk or tabletop by building this handsome desktop storage console. Designed to match the walnut writing desk (pages 256–263), this simple project converts little-used space into useful storage with five cubbies and an upper shelf. The project can be built from one sheet of walnut plywood in less than a day's time.

For companion desk plans, see pages 256–263.

Vital Statistics: Walnut Desktop Console

Type: Desk console

Overall Size: 13" D x 46" L x 18" H

Material: Walnut plywood

Joinery: Butt joints

Construction Details:

- Simple butt joints reinforced with biscuits and nails

- Plywood edges taped with iron-on veneer

Finish: Two coats of satin polyurethane varnish; match your finish to the walnut writing desk if you are building this project as a companion piece

Building Time

 Preparing Stock: 2 hours

 Layout: 1 hour

 Cutting Parts: 2–3 hours

Assembly: 2–3 hours

Finishing: 1 hour

Total: 8–10 hours

Tools You'll Use

- Combination square
- Bevel gauge
- Jigsaw
- Straightedge
- Biscuit joiner
- Clamps
- Deep clamp extenders (optional)
- Hammer
- Nailset

Shopping List

- ○ ¾" x 4' x 8' walnut plywood
- ○ Walnut edge banding
- ○ #20 biscuits
- ○ 2" finish nails
- ○ Wood glue
- ○ Finishing materials

Exploded View

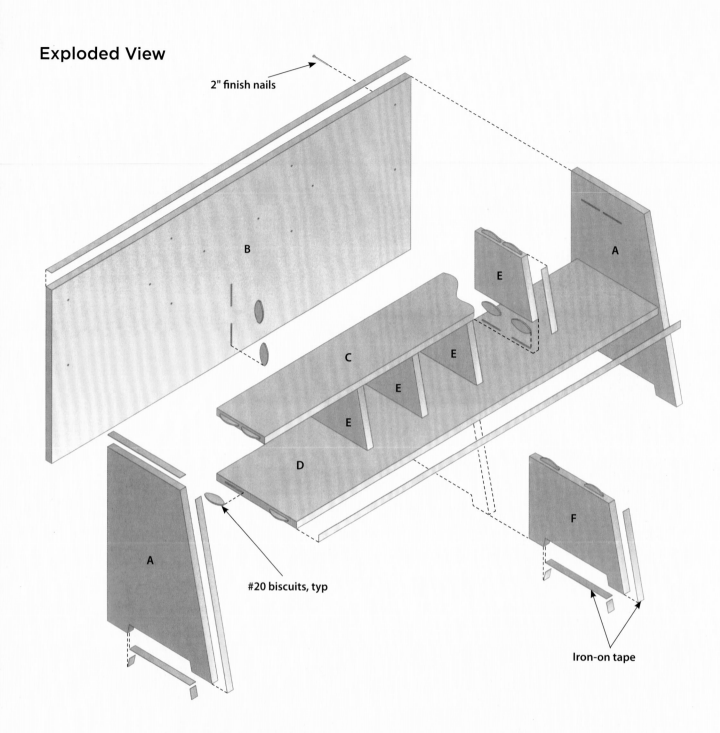

2" finish nails

B

A

E

E

C

E

E

D

#20 biscuits, typ

A

E

F

Iron-on tape

Desktop Console Cutting List

	Part	No.	Size	Material
A.	End	2	¾" x 13" x 18"	Walnut plywood
B.	Back	1	¾" x 44½" x 18"	"
C.	Top shelf	1	¾" x 6" x 44½"	"
D.	Bottom shelf	1	¾" x 10" x 44½"	"
E.	Middle divider	4	¾" x 7" x 6"	"
F.	Bottom divider	1	¾" x 11⅝" x 8½"	"

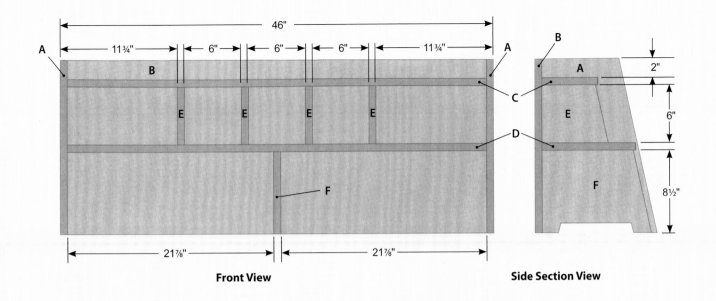

Front View

Side Section View

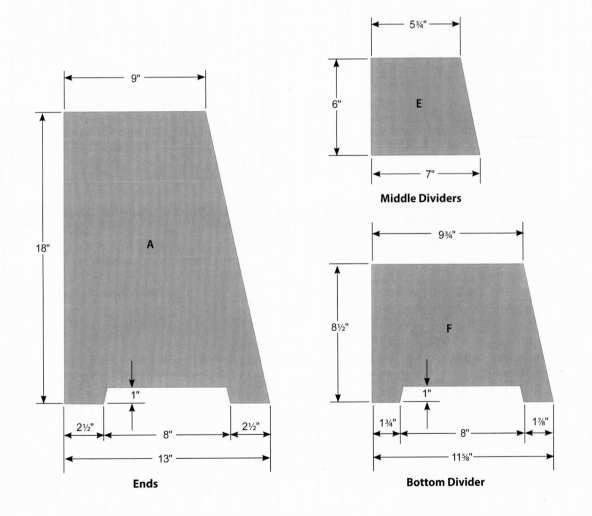

Ends

Middle Dividers

Bottom Divider

Desktop Console: Step-by-Step

Make the Ends and Dividers

1 Rip-cut and crosscut the console ends to size from ¾" walnut plywood. Along the top edge of one of the end pieces, measure over 9" from the back edge and make a reference mark. Use a straightedge to draw a line connecting this point with the lower front corner of the end piece to establish the angled profile along the console's front edge.

2 Lay out the bottom cutout on the same end piece you marked in step 1. Mark a pair of reference points 2½" from the front and back corners along the bottom edge (see Ends illustration, page 267). Set the rule on a combination square to 1" and butt the head of the square against the bottom edge of the end piece. Hold a pencil against the end of the rule and slide the square along the bottom edge to draw the top edge of the cutout. Set a bevel gauge to match the angle on the front edge of the end piece and use this setting to lay out the angled ends of the cutout (**see Photo A**).

3 Cut the console end pieces to shape. To do this, clamp the piece you just marked on top of the other end piece, aligning the edges. Gang-cut both end pieces. First, remove material in the cutout area with a jigsaw, using a fine-toothed jigsaw blade to minimize chipping the surface veneer. (These sawn edges need to be smooth; you'll cover them with walnut veneer later.) Then clamp a straightedge in place to guide the jigsaw when cutting the angled profile along the front edge of each end piece. Smooth all cut edges with sandpaper.

4 Lay out and cut the bottom divider to size. The cutout area and the angled front edge of the bottom divider should match the profile of the end pieces. Lay out the cutout on the bottom divider the same way you did for the end pieces, this time measuring in 1⅞" from the front and back corners. Use one of the console ends as a pattern for drawing the cutout and establishing the angled front edge profile. Complete the cutout and front profile cuts with a jigsaw.

5 Measure and cut four middle dividers using the Middle Dividers illustration on page 267 as a layout guide. The angled profile on the front edge of each middle divider should match those cut in the end pieces and bottom divider.

Photo A: Lay out the front angle on the end pieces by making a mark 9" out from the back along the top edge and connecting the mark with the bottom front corner. Use a bevel gauge set to this angle to lay out the ends of the bottom cutout, and a combination square to lay out the depth of the cutout. Make the cuts with a jigsaw against a straightedge guide.

Tip

The veneer tape will fit more tightly in the cutout corners if you first cut a small bevel along the ends of the tape with a sharp chisel.

6 Conceal the front and top edges of the console ends with iron-on walnut veneer edge tape (for more on applying veneer edge tape, see page 186). Apply veneer tape to the front edges of the dividers as well. Then cover each of the three edges of the bottom cutouts on the end pieces and bottom divider with veneer tape. When veneering the cutouts, apply veneer tape to the long sections first, then install the short end pieces of tape, butting the tape into the cutout corners. Trim off any overhanging tape with a chisel, edge-banding trimmer, or utility knife.

Cut the Shelves and Back

7 Cut the console back and top and bottom shelves from ¾" walnut plywood. Apply veneer tape to the top edge of the console back and the front edges of the shelves. The rest of the edges on the shelves and back can be left un-taped—they'll be concealed when you assemble the console.

Cut the Biscuit Slots

Tip

Since the desktop console bears little weight, we used #20 biscuit slots and glue for assembling all the console parts. When cutting biscuit joints, lay one part face up on the workbench and clamp the mating part flat on top of it, aligning the layout lines of both halves of the joint. With the fence set at 90°, butt the biscuit joiner base against the end of the mating part to cut the joint. Then flip the biscuit joiner around with its base flat against the face of the first part to cut the joint in the end of the mating part.

Photo B: Lay out and cut biscuit slots in all console parts. Mark the joint locations with chalk to make them easier to see. Use one of the workpieces as a straightedge to align the fence of the joiner when cutting slots on the face sides of parts. Where possible, cut both slots for a joint one after the next to keep part orientation clear.

8 Start by marking slot locations for the middle dividers on the appropriate faces of the top and bottom shelves (see Front View, page 267). With the back edges of the dividers and the shelves flush, lay out biscuit joints in these parts and cut the slots. Mark lines for the bottom divider location on the lower surface of the bottom shelf, and cut two biscuit slots in each part.

9 Mark and cut slots for biscuits that will join the shelves to the console end pieces (**see Photo B**). The bottom edge of the bottom shelf is 8½" up from the bottom edges of the end pieces. The top edge of the top shelf is 2" down from the top edges of the ends. Since the back panel fits between the end pieces, be sure to account for the thickness of the back panel when positioning slots for the shelves.

Photo C: Protect the mating surfaces with tape, and then brush on finish to all the interior surfaces. The cubbyholes in the console would be difficult to finish thoroughly if the console were assembled first.

10 Cut slots for a joint along the back edge of the bottom divider and the front face of the back panel. Assemble all console parts without glue to check the fit, then disassemble the parts for further preparation.

Finish Interior Surfaces

11 Once the console is glued up, the cubby areas and inside corners are difficult to reach for applying finish. Instead, finish all the interior surfaces of the console parts before assembly. Sand smooth all surfaces that will face into the console and ease sharp edges where the veneer edging meets the faces of the plywood. Remove all sanding dust with a brush, followed by a careful rub-down with a tack cloth. Cover joint areas with strips of masking tape to keep the biscuit slots dry. Lay out all parts and apply your desired finish, being careful not to get finish on any exposed, biscuited ends (**see Photo C**). We used two coats of satin polyurethane varnish.

Assemble the Console

12 Start by gluing and clamping the middle dividers between the top and bottom shelves (**see Photo D**). Use cauls or deep clamp extenders to spread clamping pressure evenly across the biscuit joints. Be sure the back edges of the dividers and shelves remain flush as you clamp. The biscuit slots will allow some play between the parts if you need to make minor adjustments. Use wax paper to protect the benchtop and keep from inadvertently gluing the console assembly to the bench. Clean up any excess glue with a damp rag.

13 Use a combination square and tape measure to draw lines on the outside of the back panel, marking locations for the dividers and shelves. These lines will guide you when nailing the back in place. The centerline of the top shelf is 2⅜" from the top edge of the back. The centerline of the bottom shelf is 8⅞" from the bottom edge.

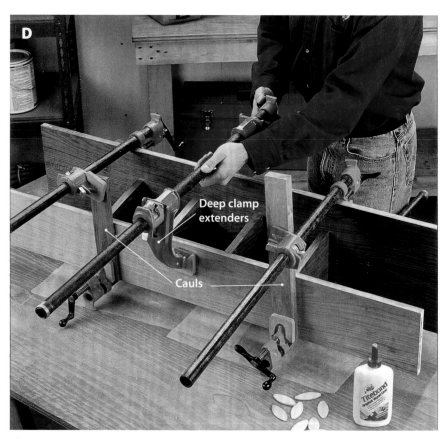

Photo D: Glue the middle dividers to the top and bottom shelves. Use scrap-wood cauls or deep clamp extenders to apply even clamping pressure across the joints. We used both homemade scrap-wood cauls and store-bought clamp extenders.

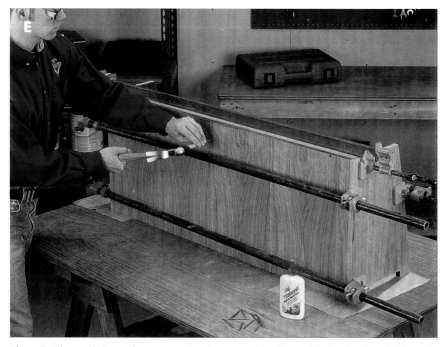

Photo E: Glue and clamp the ends to the shelves and back panel. Drive finish nails through the back into the shelves and the middle dividers, and use a nail set to sink the nail heads below the surface.

14 After the glue has dried on the first assembly, glue and clamp the shelves and back in place between the ends, keeping the top edge of the back flush with the top edges of the ends. Drive finish nails through the back into the shelves and middle dividers, using the reference lines you drew to position the nails (**see Photo E**). Set the nail heads below the surface with a nail set.

15 Attach the bottom divider to the back panel and the underside of the bottom shelf using biscuits. Clamp the bottom divider in place (**see Photo F**).

16 Sand the outside surfaces of the console, and ease any remaining sharp edges. Apply finish to all remaining surfaces.

Attaching the Desktop Console to a Desk

If you are building this desk console as a companion piece to the writing desk featured on pages 256–263, here's a tip for securing the console to the desktop: Attach 1½"-long brass mending plates, available at any hardware store, between the back of the console's back panel and the back edge of the desktop. Mending plates will keep the console from shifting, should the desk get jarred or moved (**see Photo G**).

Photo F: Attach the bottom divider to the bottom shelf and the back with glue and biscuits, clamping it in place. Use scrap blocks to pad the clamps. Make sure the front of the bottom divider is square to the shelf.

Photo G: Use 1½"-long brass mending plates to attach the console to the desktop to keep it from shifting.

INDEX

Note: Page numbers in *italics* indicate projects and jigs/aids to build. Page numbers in parentheses indicate intermittent references.

Photo Credits:

Page 11 (table saw blade): Azzaro/Shutterstock

Page 19 (top): Milwaukee Tool;

Pages 23, 138, 146 (top), and 148 (top): Delta International Machinery Corp.;

Page 25: Laguna Tools;

Pages 31–33: ©SawStop LLC, used with permission, above;

Page 35 (first-aid kit): omphoto/Shutterstock

Page 35 (telephone): immfocus studio/Shutterstock

Page 35 (fire hydrant): dcwcreations/Shutterstock

Page 104: Larisa Rudenko/Shutterstock

Pages 136 and 146 (bottom): HTC Products, Inc.;

Page 139: Incra Precision Tools;

Page 176: APA-The Engineered Wood Association;

Pages 175, 178, and 179: Willamette Industries, Inc.